ETHNOZOOARCHAEOLOGY

ETHNOZOOARCHAEOLOGY
The present and past of human–animal relationships

edited by

Umberto Albarella and Angela Trentacoste

OXBOW BOOKS
Oxford and Oakville

Published by
Oxbow Books, Oxford, UK

© Oxbow Books and the individual authors, 2011

ISBN 978-1-84217-997-0

This book is available direct from:

Oxbow Books, Oxford, UK
(Phone: 01865-241249; Fax: 01865-794449)

and

The David Brown Book Company
PO Box 511, Oakville, CT 06779, USA
(Phone: 860-945-9329; Fax: 860-945-9468)

or from our website

www.oxbowbooks.com

Front cover: Cattle ploughing in Peru. (Photo by Niels Johannsen)
Back cover: Fishing arrangements in the small open boats
(Drawn by Ruby N. Cerón-Carrasco based on a painting by the Icelandic artist Finnur Jonsson)

A CIP record for this book is available from the British Library

Library of Congress Cataloging-in-Publication Data

Ethnozooarchaeology : the present and past of human-animal relationships / edited by Umberto Albarella and Angela Trentacoste.
    p. cm.
  Includes bibliographical references.
  ISBN 978-1-84217-997-0
  1. Ethnoarchaeology. 2. Animal remains (Archaeology) 3. Human-animal relationships--History. 4. Ethnology. 5. Human-animal relationships. 6. Archaeology--Methodology. I. Albarella, Umberto. II. Trentacoste, Angela.
  CC79.E85E87 2011
  304.2--dc22
                                        2011009345

Printed and bound in Great Britain by
Short Run Press, Exeter

Contents

Preface	vi
List of Contributors	vii

PART I: INTRODUCTION AND METHODS

1. Ethnozooarchaeology and the power of analogy *(Umberto Albarella)*	1
2. A dog is for hunting *(Karen D. Lupo)*	4
3. Past and present strategies for draught exploitation of cattle *(Niels Johannsen)*	13
4. Animal dung: Rich ethnographic records, poor archaeozoological evidence *(Marta Moreno-García and Carlos M. Pimenta)*	20
5. Folk taxonomies and human–animal relations: The Early Neolithic in the Polish Lowlands *(Arkadiusz Marciniak)*	29

PART II: FISHING, HUNTING AND FORAGING

6. The historical use of terrestrial vertebrates in the Selva Region (Chiapas, México) *(Eduardo Corona-M and Patricia Enríquez Vázquez)*	41
7. Pacific Ocean fishing traditions: Subsistence, beliefs, ecology, and households *(Jean L. Hudson)*	49
8. The ethnography of fishing in Scotland and its contribution to icthyoarchaeological analysis in this region *(Ruby N. Cerón-Carrasco)*	58
9. Contemporary subsistence and foodways in the Lau Islands of Fiji: An ethnoarchaeological study of non-optimal foraging and irrational economics *(Sharyn Jones)*	73
10. Ethnozooarchaeology of the Mani (Orang Asli) of Trang Province, Southern Thailand: A preliminary result of faunal analysis at Sakai Cave *(Hitomi Hongo and Prasit Auetrakulvit)*	82

PART III: FOOD PREPARATION AND CONSUMPTION

11. An ethnoarchaeological study of marine coastal fish butchery in Pakistan *(William R. Belcher)*	93
12. Ethnozooarchaeology of butchering practices in the Mahas Region, Sudan *(Elizabeth R. Arnold and Diane Lyons)*	105

PART IV: HUSBANDRY AND HERDING

13. Social principles of Andean camelid pastoralism and archaeological interpretations *(Penelope Dransart)*	123
14. Incidence and causes of calf mortality in Maasai herds: Implications for zooarchaeological interpretation *(Kathleen Ryan and Paul Nkuo Kunoni)*	131
15. A week on the plateau: Pig husbandry, mobility and resource exploitation in central Sardinia *(Umberto Albarella, Filippo Manconi and Angela Trentacoste)*	143
16. A pig fed by hand is worth two in the bush: Ethnoarchaeology of pig husbandry in Greece and its archaeological implications *(Paul Halstead and Valasia Isaakidou)*	160

Preface

This book was originally conceived as a session that I organised as part of the 10th conference of the International Council of Archaeozoology (ICAZ) in Mexico City, on August 23rd–28th, 2006. That session included 16 oral presentations, most of which are included in their written version in this volume. For a variety of reasons a few contributions were lost on the way, but at the same time a few that were not part of the original session were added. This means that, entirely coincidentally, the number of chapters (16) is the same as the oral presentations given at the original session. All chapters in this book have been peer reviewed and I am very grateful to the many colleagues around the world who helped us in this important process, as well as to all contributors who, invariably, took referees' and editorial comments in the right spirit of constructive criticism. It is as a consequence of such positive attitude that – I believe – the quality of the book has been substantially improved. It is very unfortunate that the preparation of this book has taken so long – almost four years – and I must express my deepest gratitude to all contributors for their forbearance and trust in the fact that the editors would finally produce. I am embarrassed particularly towards those authors who handed in their papers within the deadline that we had originally agreed, and humbled by their loyalty to this project. In these last six months I have been determined to avoid that this project would drag for much longer and in this I have benefited from the fundamental help of Angela Trentacoste who, with her painstakingly thorough work and admirable designing skills, fully deserves to be acknowledged as a co-editor of the book. I would also like to express my gratitude to William Belcher who, in addition to writing an excellent chapter, also generously helped me with some other editorial aspects. My thanks also go to the Royal Society, which funded my trip to Mexico City, to Anne Pike-Tay, Greg Monks and Sue Stallibrass, who helped me to chair the session, and to the organizers of ICAZ 2006 – Joaquin Arroyo-Cabrales and the late Oscar Polaco – without whom this project would have never taken off. Finally, I must express my gratitude to Oxbow Books, with whom I had just completed the 'adventure' of the full publication of the ICAZ 2002 proceedings, for keeping faith in my work and agreeing to publish this book. Apart from the inevitable frustrations of not progressing as rapidly as I would have liked, the preparation of this book has, however, been an illuminating experience, which has allowed me to familiarize myself with a wealth of fascinating ethnozooarchaeological work carried out around the world, and given me the opportunity to work closely with a formidable group of esteemed and understanding colleagues. If this book will manage to preserve some of the excitement and interest that accompanied the session in Mexico City it will have achieved its objective. After the session had just been concluded, a young Danish colleague told me that she had been so enthused by what she had heard that she had decided to become herself involved in ethnozooarchaeology. She was looking forward to putting her ideas into practice, operating alongside her anthropologist fiancée who worked in Africa. It is terribly sad that after just a year following that conversation her life was tragically cut short, preventing her from the opportunity to fulfil her dream. This book is dedicated to her memory, enthusiasm and joy for life. We miss you Stine.

Umberto Albarella, 22nd June 2010

List of Contributors

UMBERTO ALBARELLA
Department of Archaeology, University of Sheffield
Northgate House
West Street
Sheffield S1 4ET
England (UK)
u.albarella@sheffield.ac.uk

ELIZABETH ARNOLD
Department of Anthropology
1158 Au Sable Hall
Grand Valley State University
Allendale, MI 49401
USA
arnoleli@gvsu.edu

PRASIT AUETRAKULVIT
Department of Archaeology
Faculty of Archaeology
Silpakorn University
31 Na Pralan Rd, Phranakorn
Bangkok 10200
Thailand
pratieng@yahoo.com

WILLIAM BELCHER
Central Identification Laboratory
Joint POW/MIA Accounting Command
310 Worchester Avenue, Bldg. 45
Hickam AFB, HI 96853-5530
USA
wbelcher@msn.com

RUBY CERÓN-CARRASCO
School of History, Classics and Archaeology
University of Edinburgh
Room 1M.30, Doorway 4, Teviot Place
Edinburgh EH8 9AG
Scotland (UK)
&
Historic Scotland
Collections Unit
Understanding and Access
Properties in Care
Longmore House
Edinburgh EH8 1SH
Scotland (UK)
ruby.ceron-carrasco@ed.ac.uk

EDUARDO CORONA-M
Centro INAH Morelos
Matamoros 14, Col. Acapantzingo
Cuernavaca, Morelos 62440
México
ecoroma09@gmail.com

PENELOPE DRANSART
School of Archaeology, History & Anthropology
University of Wales Trinity St David
Lampeter SA48 7ED
Ceredigion
Wales (UK)
p.dransart@tsd.ac.uk

PATRICIA ENRÍQUEZ VÁZQUEZ
Centro INAH Morelos
Matamoros 14, Col. Acapantzingo
Cuernavaca, Morelos 62440
México

PAUL HALSTEAD
Department of Archaeology, University of Sheffield
Northgate House
West Street
Sheffield S1 4ET
England (UK)
p.halstead@shef.ac.uk

HITOMI HONGO
School of Advanced Sciences
Graduate University for Advanced Studies
Hayama, Miura
Kanagawa 240-0193
Japan
hongouhm@soken.ac.jp

JEAN HUDSON
Department of Anthropology
Sabin Hall 290
3413 N. Downer Avenue
Milwaukee, WI 53211
USA
jhudson@uwm.edu

VALASIA ISAAKIDOU
Department of Archaeology, University of Sheffield
Northgate House
West Street
Sheffield S1 4ET
England (UK)
v.isaakidou@shef.ac.uk

SHARYN JONES
Department of Anthropology
University of Alabama at Birmingham
Heritage Hall, Rm. # 315
Birmingham, AL 35294-3350
USA
sharynj@uab.edu

NIELS JOHANNSEN
Section for Prehistoric Archaeology
University of Aarhus
Moesgaard
DK-8270 Højbjerg
Denmark
niels.johannsen@hum.au.dk

KAREN LUPO
Department of Anthropology
Washington State University
P.O. Box 644910
Pullman, WA 99164-4910
USA
klupo@mail.palouse.com

DIANE LYONS
Department of Archaeology
8th Floor Earth Sciences Building
University of Calgary
2500 University Drive NW
Calgary, AB T2N 1N4
Canada
dlyons@ucalgary.ca

FILIPPO MANCONI
Via Daniele Manin 25
Tempio Pausania (SS) 00129
Italy
filippo.manconi@tiscali.it

ARKADIUSZ MARCINIAK
Institute of Prehistory
University of Poznań
Św. Marcin 78
61-809 Poznań
Poland
arekmar@amu.edu.pl

MARTA MORENO-GARCÍA
Instituto de Historia
Centro de Ciencias Humanas y Sociales (CCHS), CSIC
Albasanz 26-28, Madrid 28037
Spain
marta.moreno@cchs.csic.es

PAUL NKUO KUNONI
Kitengela
Kajiado District
Kenya
polnakuo@yahoo.com

CARLOS PIMENTA
DEPA (Divisão de Estudos Patrimoniais e Arqueociências)
IGESPAR, IP, Laboratório de Arqueozoologia
Rua da Bica do Marquês, nº 2
Lisboa 1349-021
Portugal
cpimenta@igespar.pt

KATHLEEN RYAN
African Section
University of Pennsylvania
Museum of Archaeology and Anthropology
3260 South Street
Philadelphia, PA 19104-6324
USA
kryan@sas.upenn.edu

ANGELA TRENTACOSTE
Department of Archaeology, University of Sheffield
Northgate House
West Street
Sheffield S1 4ET
England (UK)
a.trentacoste@sheffield.ac.uk

Introduction and Methods

Introduction and Methods

1. Ethnozooarchaeology and the power of analogy

Umberto Albarella

Apparently the term 'ethnoarchaeology' was first coined in 1900 by the American zoologist and anthropologist Jesse Fewkes, who regarded its practitioner to be somebody who could bring "as preparation for his work an intensive knowledge of the present life" (David and Kramer 2001). Although this definition may be regarded as far too vague by contemporary archaeologists, the concept that it expresses has been instrumental to the genesis of this book. My ambition to become more involved with the world of ethnography does indeed derive from the awareness of 'an intensive *ignorance* of the present life' – at least the kind of life that is most relevant to archaeological, and more specifically zooarchaeological, interpretation. Whatever the value of ethnography to archaeological interpretation, I have invariably interpreted my own ethnographic work as a training session aimed towards that utopian dream of 'intensive knowledge'. Beyond any theoretical concern for the usefulness of ethnographic analogy, I simply felt that I was interpreting phenomena, such as husbandry, herding and hunting, with which I had no direct experience, and I felt increasingly uneasy about this. If this sentiment represents the reason for my involvement in the world of ethnoarchaeology, of such need is this present book the rather obvious consequence. The complexity and diversity of the patterns of human behaviour are such that no single individual can possibly cover the study of their full ranges; a worldwide view necessarily requires teamwork and the contributions of a diversity of researchers and approaches. In this respect this book can be interpreted as a collection of field-based training sessions, in which the participants describe their experiences for the benefit of others (and each other).

As David and Kramer (2001, 2) have pointed out, "[e]thnoarchaeology is neither a theory nor a method, but a research strategy". This is an important concept to bear in mind, as it explains the great and healthy diversity of theoretical and methodological approaches to ethnoarchaeology which also characterise this book. This ethnography-based strategy can be regarded to belong to the more general category of 'actualistic studies' (David and Kramer 2001, 13), which also includes other important investigations, such as those generally classified as 'experimental archaeology'. Unlike the experimental archaeologist, however, the ethnoarchaeologist is an observer, albeit generally a proactive one, rather than a direct producer of evidence. Ethnoarchaeologists' active participation in present-day life, for instance by discussion/conversations or experience-sharing with members of the societies that are being investigated, inevitably leads them to seek also a historical perspective concerning how those societies behaved in the past – either through oral accounts or written documentation. This directs the researcher to a strand outside actualistic studies and move towards the field more properly defined as 'ethnohistory'. Although this categorization is useful, in reality the distinction between ethnoarchaeology and ethnohistory is often blurred, as many contributions to this book also prove. We must therefore consider that ethnoarchaeological studies very often offer a diachronic, rather than just synchronic, perspective, the length of which is very variable – ranging from years to centuries. This generates a potential continuity between archaeology and ethnoarchaeology, of which an excellent example is provided – in this volume – by the chapter by Hongo and Auetrakulvit, who apply archaeological methods to investigate a contemporary society. This diachronic perspective also addresses the criticism of the use of ethnographic parallels raised by Spriggs (2008). He laments the fact that European prehistory is unduly interpreted on the basis of Pacific analogues, which tend to ignore the history and evolutionary mechanisms of contemporary Pacific societies, as well as the impact caused on them by colonialist rule and interference. It is a fair criticism; however, it applies to the way in which ethnography is used, rather than the concept of ethnoarchaeological investigation as a whole.

Unlike ethnography, in ethnoarchaeology contemporary societies tend to represent part of a means rather than an aim. The means is to accrue evidence from modern societies that can illuminate archaeological interpretation, and as such this process inevitably brings about the issue of analogical comparison between the present and the past, or, in the case of ethnohistory, between the recent and more distant pasts. Here is not the place to begin an extensive discussion of the much debated and controversial concept of 'analogy', but a few brief considerations may help in introducing some of the interpretive dilemmas that characterise most case studies presented in this book. Doubts have often been raised on the use of ethnographic analogy as a useful heuristic tool (*e.g.* Tilley 1999; Holtorf 2000), but at the same time emphasis has been placed on the fact that "archaeologists draw upon their lives and upon everything they have read, heard about or seen in the search for possible analogies to the fragmentary remains they seek to interpret" (David and Kramer 2001, 1). In other words, if we avoid using observations of contemporary societies for archaeological interpretations, we are just left with our personal experiences which, in turn, can only be used analogically for the interpretation of the past. We cannot directly observe the past, and any attempt to improve its understanding is based on comparative models, whether they are drawn from ethnographic observations or not. This led Hodder (1982, 9) to claim that "all archaeology is based on analogy".

Conversely, Tilley (1999) believes that other heuristic tools such as metaphors and metonymies can in fact also play a role in archaeological interpretation, though he merely regards them as other forms of analogy. Holtorf (2000, 166), however, questions this view and goes further by claiming that "analogies reduce uncertainty and complexity by proposing sameness". Consequently he proposes various additional forms of archaeological interpretations, ranging from jigsaw puzzles to hypermedia. Although I am prepared to accept that it would be self-limiting not to consider the possible application of a variety of other tools of investigation in archaeology, I still do not find Holtorf's dismissal of analogy as persuasive. The reason is probably associated with a fundamental difference in the interpretation of the nature of archaeological investigation. While discussing the approach he used in his PhD dissertation, Holtorf (2000, 166) mentions that he "tried not so much to reconstruct what once was, but to make sense of the past from a viewpoint of today […] As in advertising [he] wanted to stimulate the imagination, make sense and persuade by evocation and provocation, rather than by rational convincing". This typical post-modernist approach may, I suspect, find limited sympathy in the work of many ethnoarchaeologists, including at least some of those contributing to this book. Although any attempts to understand the past will inevitably be filtered through the perception of contemporary enquiry, I do believe that 'rational convincing' still has an important role to play, and it is as part of this goal that analogy can represent a useful tool of investigation.

Holtorf's criticism is in fact probably better applied to the use of ethnographic models that are over-imposed on the past, rather than simply any analogical application. Ethnographic models generally combine many complex relationships between different elements of the human society as well as different components of the human ecosystem. To conceive even only the possibility that these could wholly be replicated in the lifestyle of past societies seems naïve and evokes the kind of 'sameness' approach criticised by Holtorf. The days of almost obsessive model-building in ethnoarchaeology seem, however, to be over and you will hardly find any example of this practice in this book. Here many different methodological approaches are presented, but they tend to be open-ended, avoiding providing rigid analogical correlates of the type advocated by Roux (2007).

This book focuses on the human–animal relationship aspects of the ethnoarchaeological 'research strategy'. Its title – ethnozooarchaeology – aims to introduce a term that has so far minimally been used in the academic literature. A search of the 'web' carried out in 2006, at the time of the original presentation of the conference session that has led to the production of this book, revealed only two mentions of the word. Four years on, in 2010, the 'web' includes seven references to 'ethnozooarchaeology' – excluding those referring to this book – which does not exactly represent a rapid or substantial spread in popularity. There are probably good reasons why the word is not widely used, but we have been keen in putting it forward, not with the aim of creating a new sub-discipline, but rather because we wanted to provoke reflection on some key aspects of zooarchaeological research, which would benefit from emphasizing their links with ethnoarchaeological studies. However obvious this may seem, it is particularly important that zooarchaeologists do not forget that they deal with remains of what once were living creatures. Bones may end up being treated by zooarchaeologists as purely inanimate objects – almost like stones, but their interpretation requires an understanding of the animals and their life cycles, of which ethnoarchaeological observations may represent a healthy reminder.

There is another important and rather thorny aspect in which an ethnozooarchaeological approach can help in appropriately approaching the study of animal remains from archaeological sites. This concerns the artificial dichotomy between an 'ecological/economic' approach on the one hand and a 'social/cultural' one on the other, which seems to afflict much of archaeological interpretation. Ethnoarchaeological analysis clearly indicates that this separation is baseless, as human–animal relationships cover all aspects of human behaviour. The issue of the distinction between 'environmental' and 'cultural' archaeologists and the consequent difficult integration of different strands of analysis does not affect at all ethnography and indeed ethnoarchaeology. Ethnoarchaeological research on human–animal relationships naturally covers economic and ecological, as well as social aspects (*cf.* Sieff 1997; Schmitt and Lupo 2008; all contributions to this volume). 'Ethnozooarchaeology' therefore reminds us of the ludicrousness of regarding the role of zooarchaeologists as restricted to the reconstruction of palaeoenvironments and palaeoeconomies. It is a false perception deriving more from the organization of archaeology as an academic discipline than any heuristic logic (*cf.* Albarella 2001).

In placing the human–animal relationship at the centre of its ethnoarchaeological investigation, this book represents a novelty in the academic literature, but it has been preceded by a number of other volumes that – though with slightly different emphases – have provided important contributions to this field of study. Among these the most relevant is probably *From Bones to Behaviour* (Hudson 1993), which applies the two main areas of actualistic studies in archaeology – ethnoarchaeology and experimental archaeology – to the analysis of faunal remains. It is in this respect reassuring that Jean Hudson, the editor of that volume, is also a contributor to the current one, therefore creating a bridge between the two projects that encompasses almost twenty years of academic activity. Hudson's volume, however, has some defining characteristics that are not shared by this book. For instance: all contributors are American; it has a special focus on hunter-gatherer societies and taphonomic analysis; and in general the book seems to be heavily inspired by a 'new archaeology' approach. Another book, which is very much relevant to the topics discussed here, is complementary to Hudson's volume for its focus on Europe – rather than America – and pastoralism – rather than hunting (Bartosiewicz and Greenfield 1999). This latter volume provides a combination of what the editors define as archaeological, historical, ethnoarchaeological and ethnological approaches, though the distinction between these two latter areas of investigation seems to be blurred. Moving away from the literature in English I am keen in acknowledging the fact that the present book is not the first to propose the term 'ethnozooarchaeology' in its title, a primacy that must be credited to an ethnoarchaeological study of the use of birds in the far south of South America (Mameli and Estévez Escalera 2004). There is a plethora of other ethnoarchaeological works and projects that provide a very useful contribution to zooarchaeology, but the above-mentioned case studies are sufficient to indicate that this volume – despite its intended novelty – does not emerge from an intellectual vacuum.

The contributions to this book purposefully provide a broad geographic range, both in terms of origins of the researchers and the object of the research. Unfortunately the loss of some contributions from the original session has meant that some of the geographic coverage has gone amiss, but we still have a rather even spread of chapters by researchers based in America and Europe, and also one from Asia. In total 11 different countries are represented. The extent of the research projects is even wider, with all main continents represented. Excluding this introduction, the remaining fifteen chapters are based on research carried out in Africa (Lupo, Moreno-García and Pimenta, Arnold and Lyons, Ryan and Nkuo Kunoni), Europe (Marciniak, Cerón-Carrasco, Albarella *et al.*, Halstead and Isaakidou), Asia (Hongo and Auetrakulvit, Belcher), North America (Corona-M and Enríquez Vázquez), South America (Dransart, Hudson) and Oceania (Jones, Hudson). In addition, the contribution by Johannsen is worldwide, touching on evidence from Europe, South America, Africa and Asia. Thematically the book also provides a diversity of perspectives that we have tried to classify into the more methodologically oriented papers, and those dealing with subsistence practices (fishing, foraging, hunting), food preparation and consumption, and finally, husbandry and herding. Despite the diversity presented in this book the range of human–animal relationships is such that only a fraction of it can here be represented. I hope that these examples, rather than generating ethnographic models that will acritically be applied to archaeological interpretation, will provide useful food for thought to those archaeologists who look at the present and the past with equal curiosity and investigative zeal.

References

Albarella, U. (2001) Exploring the real Nature of environmental archaeology. In U. Albarella (ed.) *Environmental Archaeology: Meaning and Purpose*, 3–13. Dordrecht, Kluwer Academic Publishers.

Bartosiewicz, L. and Greenfield, H. J. (1999) *Transhumant pastoralism in southern Europe. Recent perspectives from archaeology, history and ethnology*. Budapest, Archaeolingua.

David, N. and Kramer, C. (2001) *Ethnoarchaeology in action*. Cambridge, Cambridge University Press.

Hodder, I. (1982) *The present past*. London, Batsford.

Holtorf, C. (2000) Making sense of the past beyond analogies. In A. Gramsch (ed.) *Vergleichen als archäologische Methode. Analogien in den Archäologien*, 165–175. BAR International Series 825. Oxford, Archaeopress.

Hudson, J. (ed.) (1993) *From bones to behaviour. Ethnoarchaeological and experimental contributions to the interpretation of faunal remains*. Occasional Paper No. 21. Carbondale, Center for Archaeological Investigations, Southern Illinois University at Carbondale.

Mameli, L. and Estévez Escalera, J. (2004) *Etnoarqueozoología de aves: el ejemplo del extremo sur Americano*. Madrid, Consejo Superior de Investigaciones Científicas.

Roux, V. (2007) Ethnoarchaeology: a non historical science of reference necessary for interpreting the past. *Journal of Archaeological Method and Theory* 14(2), 153–178.

Schmitt, D. N. and Lupo, K. D. (2008) Do faunal remains reflect socioeconomic status? An ethnoarchaeological study among Central African farmers in the northern Congo Basin. *Journal of Anthropological Archaeology* 27, 315–325.

Sieff, D. F. (1997) Herding strategies of the Datoga pastoralists of Tanzania: is household labor a limiting factor. *Human Ecology* 25(4), 519–544.

Spriggs, M. (2008) Ethnographic parallels and the denial of history. *World Archaeology* 40(4), 538–552.

Tilley, C. (1999) *Metaphor and material culture*. Oxford, Blackwell.

2. A dog is for hunting

Karen D. Lupo

While the origins and timing of dog domestication are the focus of a number of recent studies, an equally important issue concerns 'why?'. A number of researchers nominate the value of early dogs in cooperative hunts involving larger-sized prey. But prehistoric dogs spread very rapidly to many different habitats and were likely deployed in a variety of different hunting contexts. In this paper I report ethnoarchaeological data on how dogs are deployed and influence the hunting success of smaller-sized game among contemporary forest foragers in a Central African rainforest. In this context dogs play an assisting role in some, but not all, types of hunts. Finally, I discuss how differences in dog deployment strategies might be reflected in the archaeological record and influence the composition of zooarchaeological assemblages.

Keywords: ethnoarchaeology, forest foragers, dogs, central Africa, hunting technology

Introduction

Dogs (*Canis familiaris*) are the earliest and most versatile of all domesticated animals. A number of recent studies have focused on the origins and timing of dog domestication (*e.g.* Crockford 2006; Savolainen *et al.* 2002; Sundqvist *et al.* 2006; Verardi *et al.* 2006; Verginelli *et al.* 2005; Vila *et al.* 2002). Recent mitochondrial DNA analyses, for example, suggests that dogs diverged from wolves as early as 134,000 years ago but were morphologically indistinguishable from their wild progenitors until 15,000–10,000 years ago when human populations became less mobile (Vila *et al.* 2002). Two possible routes for dog domestication are implied by molecular analyses: either a single event involving one wolf population (Savolainen *et al.* 2002; Sundqvist *et al.* 2006) or multiple events in different localities with continued interbreeding between wolves and dogs in some areas (Ciucci *et al.* 2003; Tchernov and Valla 1997; Verardi *et al.* 2006; Verginelli *et al.* 2005). Molecular and archaeological evidence are not in precise agreement regarding the timing of dog domestication. The earliest archaeological evidence for identifiable domesticated dogs date to 17,000–13,000 ^{14}C years BP and were recovered from Eliseevichi I on the Central Russian Plain (Sablin and Khlopachev 2002). Several other early finds of dog remains date between 14,000 and 12,000 BP (see Crockford 2006, 95). These and other archaeological finds in the Near East, Europe and Siberia show that early dogs were morphologically distinct from wolves, but overlapped in overall body size and form (Dobney and Larson 2006; Morey 2006; Musil 1984; Olsen 1985; Turnbull and Reed 1974). By 10,000 years ago dogs are associated with human settlements in three continents (Verginelli *et al.* 2005) and some 7,000 to 4,000 years ago show morphological differentiation in some areas (Clutton-Brock 1999; Lupo and Janetski 1994). Regardless of how and when dogs became domesticated, modern dog breeds display a high degree of phenotypic plasticity and important behavioural and cognitive differences not found in their wild progenitors (Bjornerfeldt *et al.* 2006; Crockford 2006; Hare *et al.* 2002; Miklosi *et al.* 2003; Saetre *et al.* 2004).

While a great deal of research has focused on the origins and timing of domestication, an equally important question concerns 'why?'. Most discussions point to the myriad of functions served by modern dogs in contemporary and historic societies. These include their ability to hunt, herd and guard livestock and people, transport loads, clean garbage, serve as companions and symbols of power and ritual, and their consumptive utility (*e.g.* Snyder and Moore 2006). Some, but not all, of these roles are based on traits amplified by modern selective breeding and have only recently emerged (*e.g.* Morey and Aaris-Sorensen 2002). It is not clear what niche the earliest dogs and their tame progenitors filled in prehistoric societies, but most studies cite their value in the cooperative hunting of larger-sized prey. The use of early domesticated and proto-dogs in human cooperative hunts is often viewed as an extension and modification of their pre-existing predation pattern. Contemporary wild wolves acquire most (but not all) of their prey in cooperative efforts, and it is likely that prehistoric wolves behaved in a similar fashion. Cooperative hunting involving early dogs likely targeted the same prey as their wild canid progenitors. Thus, early dogs were pre-adapted to the cooperative deployment and

Fig. 2.1 Typical forest forager dog

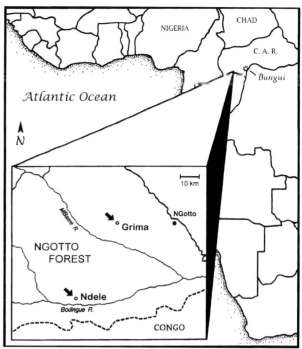

Fig. 2.2. Map showing the study area and villages

the predatory responses of specific animals. But early dogs spread very rapidly into many different habitats where they were likely deployed in many different ways and targeted a variety of different prey. A central question is how the use of dogs influenced different kinds of hunting strategies and how differences in dog deployment might be manifested in the archaeological record.

In this paper I use ethnoarchaeological data derived from Central African foragers and their dogs to explore differences in canine deployment in hunting and how these differences might be reflected in the archaeological record. The Central African dogs discussed here are ancestral to modern Basenji's, a breed recognized by professional kennel associations (Fig. 2.1). Basenji's are late arrivals in sub-Saharan Africa and possibly accompanied Bantu populations who spread east and southward some 2,000 years ago (Greyling *et al.* 2004).

Study Area

Data reported here were collected as part of an ethnoarchaeological project on hunting among contemporary Bofi and Aka forest foragers in the southwestern Central African Republic (Lupo and Schmitt 2005; 2004; 2002). These data are derived from over 238 days of observation spanning several wet and dry seasons in two different villages and a series of interviews with 10 hunters (five Bofi and five Aka) about their dogs. The study focused on hunting in the villages of Grima and Ndele, located in the N'gotto Forest Reserve (Fig. 2.2). The study village of Grima is occupied by 150 Bofi foragers, while Ndele is inhabited by 111 Aka and approximately 25 Bofi foragers.

The N'gotto Forest Reserve is located in an area characterized by tropical microenvironments including rain forests, ephemeral wetlands, and wet savannas (Bahuchet and Guillame 1982). The vegetation in this area is classified as a drier type of Guinea-Congolian rainforest (White 1983) and is especially rich in *Entandrophragma cylindricum* and *E. utile* which are highly valued by logging companies (Ngasse 2003). High annual average temperatures (around 77°F), humidity (70–90%) and precipitation characterize this area. During the wet season, mid-June to October, heavy and almost daily rains fall with monthly averages sometimes exceeding 200 mm (Hudson 1990). Considerably less precipitation falls in the dry season, December through May.

The Aka and Bofi are two related, but ethnolinguistically distinct, groups of forest foragers. Despite the differences in their language, the Bofi claim a close ancestral relationship to the Aka, and there are many material similarities between the two groups in house construction, hunting and subsistence technology. There are no differences between the Bofi and Aka in hunting technology or how dogs are deployed. The Bofi and Aka have interdependent and multidimensional relationships with settled farmers who occupy permanent villages. This relationship has economic, social and religious dimensions, but the exchange of forest products, such as meat and honey for manioc and other products, is the most visible and prominent aspect of this relationship. The relationship between foragers and settled farmers is believed to be very old and extend back some 2,000 years to when Bantu horticulturalists arrived in the area.

The Bofi and Aka occupy permanent residential camps and a series of temporary forest camps throughout the year. Permanent residential camps are maintained next to farmer villages and may be occupied for up to six months or longer by segments of the forager population. The foragers also use a series of temporary forest hunting camps as bases for procuring forest resources for trade and consumption. These camps may be occupied for up to several months by a single family or larger population aggregates.

About one half of the diet is obtained from gathering wild products and hunting animals. Gathered products include wild plants such as koko (*Gnetum africanum*), several different species of wild yam (*Dioscorea* sp.) and mushrooms (Pleurotoidea). A variety of fruits and nuts are consumed including *Trecula africana*, *Irvingia robur* and *Irvingia gabonensis*. Insects, including termites, caterpillars and butterfly pupa, and land snails comprise important collected resources. Honey from several species of honeybees and stingless bees is also highly valued (Kitanishi 1995). Hunting is considered an important activity and meat is a highly prized source of food. Meat is obtained by hunting and, on rare occasions, scavenging wild prey. The most common prey are less than 10kg in live weight and include blue duikers (*Cephalophus monticola*), giant pouched rats (*Cricetomys emini*), brushy-tailed porcupine (*Atherurus africanus*), guenon monkeys, small carnivores, reptiles and birds. Medium-sized prey (10 to 25kg) includes Bay and Peters duikers (*C. dorsalis, C. callipygus*), which are uncommon in and around Grima, but are encountered more frequently near Ndele. Larger-sized prey (>25kg), such as yellow-backed duiker (*C. silvicutor*) and river hog (*Potamocherus porcus*), are uniformly uncommon in this area. The largest traditional prey, elephant (*Loxodonta africana*), is currently rare and protected by law.

The Bofi and Aka use a wide variety of communal and individual hunting techniques to obtain prey. The best known of the communal techniques is the net hunt which involves men, women and children using hand made fibre nets (Harako 1976; Lupo and Schmitt 2002; 2004; Terashima 1983). While a variety of resources are encountered and pursued during these hunts, nets target dense but randomly distributed terrestrial prey that can be easily flushed, especially blue duikers. Individual hunting can involve one to three people and includes the use of spears, traps, snares, crossbows and hand capture. The most common prey taken with spears are medium and larger-sized duikers that are too large to be caught in nets and other animals that are difficult or dangerous to handle, such as porcupine, and small carnivores. The hand capture of prey involves the use of fire, dogs, and digging implements and is aimed at animals that are fossorial, solitary and non-aggressive, such as giant pouched rats, pangolins and tortoises. Snares made from fibre or metal cable are not a traditional hunting technique used by foragers, but the use of this technology is increasing despite the high cost of the cable. Snares are usually generic in form (*i.e.* simple noose form snare) but are scaled to the size of the animal and target a wide range of prey, especially those known to use habitual runways or trails (see also Lupo and Schmitt 2005; Noss 1995; 1998). Small traps are not very common and include devices designed to entrap prey via complete enclosure. Only two kinds of traps are used with any regularity: a small fibre purse trap largely used to obtain porcupine and rat and a woven cone trap used to procure murid rats and mice (see Lupo and Schmitt 2005). In the past, crossbow hunting with poisoned darts was used to procure arboreal animals such as monkeys, bats and birds, but now most of these animals are largely hunted with guns. Only one forager in our study sample owned a gun, which was in disrepair, but village farmers will often hire foragers to hunt and lend them their guns.

Forest Dogs

Dogs are kept by the foragers and farmers in both villages, but are not particularly numerous. Among foragers approximately 50% of the households had one or two dogs. Few households ever had more than two unless a female dog had just given birth. Dogs are generally roughly treated and puppies only slightly less so. It is not uncommon to see these animals kicked, hit or thrown out of huts. Puppies learn early to approach humans with extreme caution. On rare occasions dogs are intentionally killed because they are no longer able to hunt or have become a nuisance. One forager reported that his dog was intentionally killed by a farmer because it came too close to his house. Nevertheless, dogs and puppies are named and encouraged to thrive.

Hunters report that dogs are kept solely as hunting aids. A dog is fairly inexpensive even by local standards; an adult male can be purchased for the equivalent of half a small duiker carcass (about $1.00 US), and a female may cost a little more because of her reproductive abilities.

Bofi and Aka hunters laughed at the suggestion that dogs might be companions/friends or family members, although most acknowledged that dogs provided a valuable service in helping to obtain prey. While all hunters acknowledged that dogs kept sites clean and worked as garbage disposals, they did not cite this as an important benefit to owning a dog. No hunter cited protection in the forest as a function fulfilled by dogs; dogs often accompany foraging groups comprised only of women and children into the forest, but the dogs are taken only to help hunt for small prey that might be encountered. All hunters denied eating dog, but some reported that they knew of someone who did. On one occasion, we discovered a dog humerus fragment in a foragers garbage midden mixed with other food bones, but it was not clear how the bone got there. Both the foragers and villagers believe that dogs and other animals possess spirits and can haunt the living. Dogs are also viewed as a common physical form taken by sorcerers and witches.

While some dogs are acknowledged to be better at hunting then others, hunters reported that there were no differences between males or females in hunting ability. No attempt is made to control or regulate breeding to improve the abilities of the dogs. All dogs are trained

Fig. 2.3 Most dogs are always alert to feeding opportunities and place themselves in close proximity to food processing areas and people (especially children) who are eating.

to hunt beginning at six months of age. Young dogs are given an herbal concoction, which is put up their nose and purportedly enhances their ability to find and chase prey. Young dogs are also often taken into the forest with adult dogs and learn by observation and participation.

Dogs are provisioned and what they eat depends on their hunting success. Those that are successful are fed some of the kill in the forest, usually parts of the internal organs and blood when the animal is butchered. After the kill has been transported, prepared and consumed at camp, the dog may be given small amounts of meat and bones, usually the cranium after it has been picked clean by the foragers. If the hunt was unsuccessful, the dog is not fed meat and may not be given food of any kind. Most people said that garbage was the food of the unsuccessful hunting dog. We observed people giving dogs small bits of manioc, corn, yam and other vegetable products on several occasions even when meat was available for consumption. In camp, dogs position themselves in close proximity to food preparation areas and quickly snap up any small bits of food that are accidentally dropped. Dogs also station themselves close to very small children who may accidentally drop food (Fig. 2.3). During the wet season when hunting returns are generally poor and meat is harder to find for everyone, dogs tend to become weak and very thin.

All dogs are very lean and perpetually hungry. As such hunters occasionally loose prey when it is entirely or partially consumed by the dog before the hunter retrieves the carcass. Interestingly dogs do not feed themselves by hunting for small prey, even though they are quite capable. They usually stay in close proximity to the village or camp and, surprisingly, often show reluctance to go on hunts. On several occasions we observed hunters, dragging and even carrying the dogs into the forest to go hunting. A dog's reluctance to go hunting and procure prey may be linked to the danger the forest poses towards them. Dogs live fairly short lives in this area. While a variety of parasitic diseases can kill dogs (Nozais 2003), most deaths are attributable to other causes. Respondents reported that most dogs lived only two or three years and the oldest dog anyone could recall lived approximately five years. Most die in the forest and simply do not return from hunting trips. The most common cause of death is snakebite, followed by hunting accidents or being killed by leopards or other carnivores. Hunting accidents involves dogs accidentally being speared or hacked with a machete while attempting to flush an animal from brush or a fallen log. Moreover, some dogs were seriously injured and a few eventually died after accidentally tripping metal snares. Occasionally, an injured or sick dog dies in camp and is buried. Hunters reported that they buried dogs out of respect because they helped hunt and obtain meat.

Hunt Parameters	With a Dog	Without a Dog
Number of hunts	46	82
Proportion successful	0.43 (20)	0.34 (28)
Proportion unsuccessful[a]	0.39 (39)	0.50 (41)
Proportion rat escapes	0.13 (13)	0.16 (13)
Proportion dog eats prey[b]	0.04 (2)	0
Average time to kill prey[c]	29 mns	49.50 mns
Average time to break[d] off pursuit	13.11 mns	9.27 mns

a. Unsuccessful hunt includes all hunts abandoned either because the rat was too deeply embedded in the burrow or the burrow was deemed empty after closer inspection.
b. Cases where the dog captured the rat and completely consumed it before the hunter could reach the dog.
c. Mns=minutes.
d. Average time to break off pursuit includes cases where the burrow was abandoned (as defined above) and where the rat was pursued after running fleeing the burrow and ultimately escaped.

Table 2.1 Comparison of giant pouched rat hunting success with and without a dog

Hunt Parameters	With Dog	Without Dog
Number of hunts	35	13
Proportion successful hunts	0.45	0.46
Proportion unsuccessful hunt	0.55	0.54
Average time to kill prey[a]	44 mns	101.33 mns
Average time to break off pursuit	22 mns	3.6 mns

a: Mns=minutes.
In one case of a successful hunt, the dog consumed ½ of the porcupine before the hunter could reach the dog.

Table 2.2 Comparison of porcupine hunts with and without a dog

Hunting and Dogs

The forest dogs play important assisting roles in some, but not all, of the hunting techniques used by the Bofi and Aka (see Lupo and Schmitt 2005). In this cultural context, dogs are deployed as individuals or work in small groups (three to four dogs). The dogs are considered to be 'barkless', but are not mute and can make low barking noises. However, the dogs are not used to scare or distract prey by barking or attacking animals. Most of the prey in this area are specialized in concealment and can only flee predators over short distances (Kingdon 1997). Consequently, the dogs are largely used to flush prey from heavily vegetated thickets, logs or burrows (Hudson 1990; Lupo and Schmitt 2005).

Dogs are considered particularly useful in the hand capture of giant pouched rats. Foragers, especially women, often complained that they were unable to hunt giant pouched rats on particular days because they didn't have a dog with them. Giant pouched rats are procured by finding the entrance hole to their underground burrow and are extracted by digging or smoking them out. Dogs do not usually locate the burrows for the hunters, nor do they help dig out the animal, but they will chase and catch the rat if it escapes from the burrow. Table 2.1 compares rat hunting success with and without dogs. Dog-assisted rat hunts have a higher success rate and take less time than those without dogs. Hunters with dogs abandoned fewer hunts than those without dogs and were more persistent in pursuing a rat after it was encountered than those without dogs. Dogs make very little difference in the proportion of hunts in which the rats escapes. Note that on at least two occasions the dog completely consumed the rat before the hunter could reach the animal.

Porcupines are hunted with spears, by hand and with small traps. Dogs often assisted in these hunts but, again, did not usually locate the porcupine for the hunter; rather the hunter identifies a possible thicket, log or hole and calls the dog over to investigate. The dog assists by chasing the animal after it is located. Table 2.2 shows that the vast majority of the porcupine hunts were dog assisted; the sample of hunts without dogs is quite small. As shown, the use of dogs has very little impact on hunting success (as measured by the proportion of successful hunts) or the proportion of hunts when the prey escapes. But hunters with dogs take less time to kill the porcupine than those without dogs, and hunters with dogs were more persistent than those without dogs. We only observed one instance of a successful hunt where the dog consumed half the porcupine before the hunter could reach the animal.

Dogs were less useful on communal net hunts even though they were frequently present (also see Turnbull 1965). Hunters reported that dogs often help chase prey into the net and catch the animal, but sometimes the dog actually chased the animals the wrong way (*i.e.* out of the nets). Table 2.3 shows that dogs have a less appreciable

Hunt Parameters	With Dog	Without Dog
Number of hunts	38	8
Proportion successful hunt[a]	100	100
Proportion unsuccessful hunt	0	0
Average hunt length[b]	277 mns	282 mns
Average number prey taken	4.83	6.87
Average number of prey escapes	3.2	2.25
Average number of dogs per hunt	2.23	0

a: Successful net hunts are defined as those where at least one animal was taken in the nets. On hunts where dogs were present, the dogs may or may not have directly assisted in these captures.
b: Mns= minutes.
There were no known cases in which the dog consumed the prey before the hunter reached the animal.

Table 2.3 Comparisons of net hunts with and without dogs

influence on net hunting success. Net hunts with dogs tend to be slightly shorter than those without dogs, and net hunts without dogs actually took more prey on average than those with dogs. Slightly more animals escaped the nets on hunts with dogs than those without dogs. The lack of influence that dogs have on net hunting success is not surprising. Previous studies of forest foragers show that net hunting success depends on the number of nets and human participants rather than the number of dogs (Harako 1976; Lupo and Schmitt 2002; 2004; Terashima 1983).

Given their usefulness on other types of hunts, one might question why dogs are present on net hunts. Net hunts in this context are best viewed as general hunting opportunities rather than single task events (Lupo and Schmitt 2002; 2005). People who go on net hunts frequently pursue other hunting opportunities as they arise, such as chasing porcupines or hunting rats. People on net hunts frequently encounter medium and larger-sized prey that is too large to be caught in the nets and must be pursued on foot with spears and dogs. Dogs are useful on these occasions for their ability to track the wounded animal through the forest. As an added advantage, if a dog happens to help another person on the net hunt obtain prey, the owner of the dog is entitled to a share of the bounty.

In general, forest dogs play only a minor assisting role in hunting success among forest foragers. The use of dogs increases the success rate of hunting certain kinds of prey, but they have a less appreciable affect on cooperative net hunts. Unfortunately, the data set for other types of hunts, such as spear hunting, is quite limited and precludes the possibility for examining how dogs influence the success of these types of hunts. Nevertheless, the modest increase in hunting success associated with the use of dogs, the small numbers of animals owned by people and the relative lack of investment in them suggests dogs are not as highly valued as hunting tools as they might be in other ecological contexts.

Osteological and Archaeological Consequences

Although forager dog burials or skeletons were not examined as part of this study, it is reasonable to hypothesize how some of the circumstances described here might be manifested in different kinds of osteological and archaeological evidence. First, because many dogs die in the forest, only a small number of dog burials involving a fairly young population should occur at residential camps. In the Central African case, dog mortality is high from natural predators and hunting techniques that involve a close proximity between hunter, weapon and dog. Depending on the prey, hunting techniques and method of deployment, hunting can be dangerous even when the prey is not large-sized or particularly aggressive. Ikeya (1994) reports a similar pattern in dog mortality among the San in the central Kalahari. A high mortality resulting from hunting accidents might also explain why so few early dog burials are found in archaeological contexts.

Second, dog burials will likely be found in close proximity to human burials and living spaces. In the context described here, people lack formalized cemeteries and the dead are often interred behind huts or living areas; dogs are often buried in the same general location resulting in close spatial proximity between dog and human graves. However, it is important to note that in this context the proximity of burial location to the house and other family members does not reflect a similarly close position of the dog within the family.

Third, the presence of hunting related injuries on skeletons are likely reflect the range of hunting activities pursued by the dog. In this context, injured lower forelimbs are common and sometimes involve breaks through the radius/ulna inflicted by machetes or knives. Another common injury involved missing forepaws from being caught in metal snares. This usually involved paws being dismembered at or just below the carpals. While these injuries were inflicted with metal tools, it is likely that similar hunting injuries could result from stone tools whenever the hunt results in a close proximity between the dog and hunter. Other types of injuries are inflicted by people who intentionally hit the dogs with sticks, riffle butts and knives when the dogs got too close to the animal being butchered, refused to release a carcass or simply got in the way. These injuries appear to represent serious superficial wounds, but it is possible that underlying fractures occurred on the nasal, orbital, maxilla and frontal areas. Since the dogs often survived these injuries, the damage would possibly be reflected by bone regrowth on the underlying bone.

Finally, seasonal variability in the availability of meat and differences in hunting ability mean that dogs are likely to have a high vegetable component in their diet, which should be identifiable via stable isotope analysis (*e.g.* Clutton-Brock and Hammond 1994; Clutton-Brock and Noe-Nygaard 1990). Ethnographic evidence from other parts of Africa show that hunting dogs are intentionally fed diets high in vegetable components rather than meat, based in part on the belief that this will make them better hunting dogs (*e.g.* Bohannan and Bohannan 1966). Evidence for malnutrition in the form of enamel hypoplasia, poor bone formation and Harris lines should also be apparent in the Central African dog populations discussed here. Interestingly, the degree and severity of malnutrition suffered by the dogs can also serve as a litmus test for evaluating resource stress among human populations. In the case example discussed here, dogs are considered quite expendable and when food is in short supply the dogs simply go without. Consequently during periods of long-term stress (on an annual scale or longer), one might expect the immediate demise of dog populations, while human populations persist.

Conclusions

This ethnoarchaeological example is largely characterized by single dog deployment in the acquisition of small prey. In this context, the use of dogs increased hunting success in only a few instances, yet dogs are viewed as valuable albeit highly expendable hunting aids. But ethnographic and historic data within continental Africa and other parts of the world show that dogs are highly versatile in 'prey target', can be deployed in a variety of different ways, and their value and use is highly variable (*e.g.* Ikeya 1994; Nobayashi 2006; Snyder and Moore 2006). Understanding the differences in how contemporary dogs are deployed and influence hunting strategies can shed light on how and why early and proto-dogs spread so rapidly among prehistoric populations. A related and equally important issue concerns how different deployment strategies influence the archaeological record. While the influence of dogs (and related canids) on bone survivorship and taphonomy is fairly well known (*e.g.* Hudson 1993), the impact of different dog deployment strategies on human behaviour as manifested in zooarchaeological assemblages remains underexplored. One might hypothesize, for example, that increases in the efficiency of one type of hunt resulting from the use of dogs may impact the diet by narrowing the range of species taken by human hunters. Over time, the increased use of dogs in specific hunts could result in the localized depression of certain prey and be manifested in the composition of zooarchaeological assemblages. Recent research among contemporary hunters using dogs points to other less obvious ways that the deployment of dogs might influence the taxonomic composition and mortality profiles of zooarchaeological assemblages. For example,

research among the San in the central Kalahari shows that a shift from single dog deployment to hunting with packs resulted in a concomitant change in the range of target prey (Ikeya 1994). Similarly, Nobayashi (2006) found that dog assisted hunts among traditional hunters in Taiwan produced differences in the age structure of wild boars (*Sus scrofa taivanus*) in comparison to other hunting techniques that did not use dogs, such as snares. Dogs assisted hunts take more adult male wild boars than naturally found in the wild boar population. Identification of prehistoric patterns of dog deployment in hunting may also reveal important information about past ecological circumstances. Recent research conducted by Ruusila and Pesonen (2004) found greater success among Finnish hunting groups that used dogs as opposed to those that did not use dogs to hunt moose (*Alces alces*). The benefits of using dogs on hunts were most marked when moose densities declined, making the prey harder to find and smaller-sized human hunting groups needed. These results may point to the ecological circumstances under which early dogs became useful partners in cooperative hunts.

Increased ethnoarchaeological emphasis on how dogs work and are used by hunters in different contexts may help clarify the impact of early dogs and proto-dogs on human populations. Ultimately, additional ethnoarchaeological studies targeting dog–human interactions may help us better understand why domesticated dogs spread so widely and rapidly among prehistoric human populations and endure as companions to this day.

Acknowledgements

I thank Dave Schmitt, Barry Hewlett, Hillary Fouts, George Ngasse, Alain Kolet Guy, Eduard Mboula, Timothee Tikouzou, Gabi Mbera, Alain Peneloin, the Makenzi family, Chef Doko Molli and the folks at Hotel Levy's. I also thank ECOFAC, the Office of Scientific and Technological Research and the government of the Central African Republic for granting permission to conduct this research. Dave Schmitt, Dave and Kathy Johnson, Jason Fancher, C. T. Hall and Matt Landt helped collect the data used in these analyses. This research would not be possible without the kindness and patience of the Aka and Bofi foragers and farmers who live in Grima and Ndele and generously tolerated our work with good humour. This research was supported by grants from the L.S.B. Leakey Foundation and the National Science Foundation (BCS-0003988).

References

Bahuchet, S. and Guillame, H. (1982) Aka-Farmer Relations in the Northwest Congo Basin. In E. Leacock and R. Lee (eds.) *Politics and History in Band Societies*, 189–212. Cambridge, Cambridge University Press.

Bjornerfeldt, S., Webster, M. T. and Vila, C. (2006) Relaxation of selective constraints on dog mitochondrial DNA following domestication. *Genome Research* 16, 990–994.

Bohannan, P. and Bohannan, L. (1966) *Three Source Notebooks in Tiv Ethnography*. New Haven, Human Relations Area Files.

Ciucci, P., Lucchini, V., Boitani, L. and Randi, E. (2003) Dewclaws in wolves as evidence of admixed ancestry with dogs. *Canadian Journal of Zoology–Revue Canadienne de Zoologie* 81(12), 2077–2081.

Clutton-Brock, J. (1999) *A Natural History of Domesticated Mammals*. Cambridge, Cambridge University Press.

Clutton-Brock, J. and Hammond, N. (1994) Hot dogs: comestible canids in Preclassic Maya culture at Cuello Belize. *Journal of Archaeological Science* 21, 819–826.

Clutton-Brock, J. and Noe-Nygaard, N. (1990) New osteological and C-isotope evidence on Mesolithic dogs: comparisons to hunters and fishers at Star Carr, Seamer Carr and Kongemose. *Journal of Archaeological Science* 17, 643–653.

Crockford, S. J. (2006) *Rhythms of Life: Thyroid Hormone & the Origin of the Species*. Victoria, BC, Trafford Publishing.

Dobney, K. and Larson, G. (2006) Genetics and animal domestication: new windows on an elusive process. *Journal of Zoology* 269 (2), 261–271.

Greyling, L. M., Grobler, P. J., Van Der Bank, H. F. and Kotze, A. (2004) Genetic characterisation of a domestic dog (Canis familiaris) breed endemic to South African rural areas. *Acta Theriologica* 49(3), 369–382.

Harako, R. (1976) The Mbuti as hunters: A study of ecological anthropology of the Mbuti pygmies. *Kyoto University African Studies Monograph* 10, 37–99.

Hare, B., Brown, M., Williamson, C. and Tomasello, M. (2002) The domestication of social cognition in dogs. *Science* 288, 1634–1636.

Hudson, J. (1990) *Advancing Methods in Zooarchaeology: An Ethnoarchaeological Study Among the Aka Pygmies*. Ph.D. Dissertation, Ann Arbor, University of California at Santa Barbara. University Microfilms.

Hudson, J. (1993) The impacts of domestic dogs on bone in foragers camps. In J. Hudson (ed.) *From Bones to Behavior: Ethnoarchaeological and Experimental Contributions to the Interpretations of Faunal Remains*. Center for Archaeological Investigations Occasional Paper No. 21, 301–323. Carbondale, Southern Illinois University.

Ikeya, K. (1994) Hunting with dogs among the San in the Central Kalahari. *Kyoto University African Study Monographs* 15, 119–134.

Kingdon, J. (1997) *The Kingdon Field Guide to African Mammals*. London, Academic Press.

Kitanishi, K. (1995) Seasonal changes in the subsistence activities and food intake of the Aka hunter-gatherers in northeastern Congo. *Kyoto University African Study Monographs* 16, 73–118.

Lupo, K. D. and Janetski, J. (1994) Evidence of Domesticated Dogs and Some Related Canids in the Eastern Great Basin. *Journal of California and Great Basin Anthropology* 16(2), 199–220.

Lupo, K. D. and Schmitt, D. N. (2002) Upper Paleolithic net hunting, small mammal procurement and women's work effort: a view from the ethnographic and ethnoarchaeological record of the Congo Basin. *Journal of Archaeological Method and Theory* 9(2), 147–180.

Lupo, K. D. and Schmitt, D. N. (2004) Meat-sharing and the archaeological record: a preliminary test of the show-off hypothesis among Central African Bofi foragers. In G. Crothers (ed.) *Hunters and Gatherers in Theory and Archaeology*. Center for Archaeological Investigations Occasional Paper No. 31, 241–260. Carbondale, Southern Illinois University.

Lupo, K. D. and Schmitt, D. N. (2005) Small prey hunting technology and zooarchaeological measures of taxonomic diversity and abundance: Ethnoarchaeological evidence from Central African forest foragers. *Journal of Anthropological Archaeology* 24(4), 335–353.

Miklosi, A., Kubinyi, E., Topal, J., Gacsi, M., Viranyi, Z. and Csanyi, V. (2003) A simple reason for a big difference: wolves do not look back at humans, but dogs do. *Current Biology* 13(9), 763–766.

Morey, D. F. (2006) Burying key evidence: the social bond between dogs and people. *Journal of Archaeological Science* 33, 158–175.

Morey, D. and Aaris-Sorensen, K. (2002) Paleoeskimo dogs of the eastern Arctic. *Arctic* 55(1), 44–56.

Musil, R. (1984) The first known domesticated wolves in central Europe. In C. Grigson and J. Clutton-Brock (eds.) *Animals and Archaeology: 4. Husbandry in Europe*. BAR International Series 227, 23–26 Oxford, Archaeopress.

Ngasse, G. (2003) Sustainable management of tropical forests in Central Africa. In I. Amsallem, M. L. Wilkie, P. Koné and M. Ngandji (eds.) *In Search of Excellence*. Rome, Food and Agricultural organization of the United Nations (on-line version).

Nobayashi, A. (2006) An ethnoarchaeological study of chase hunting with gundogs by the aboriginal peoples of Taiwan. In L. M. Snyder and E. A. Moore (eds.) *Dogs and People in Social, Working, Economic or Symbolic Interaction*, 77–84. Oxford, Oxbow Books.

Noss, A. J. (1995) *Duikers, Cables and Nets: A Cultural Ecology of Hunting in a Central African Forest*. Unpublished Ph.D. dissertation. University of Florida, Gainesville.

Noss, A. J. (1998) The impacts of cable snare hunting on wildlife populations in the forests of the Central African Republic. *Conservation Biology* 12(2), 390–398.

Nozais, J. P. (2003) The origin and dispersal of human parasitic diseases in the Old World (Africa, Europe and Madagascar). *Memorias Do Instituto Oswaldo Cruz* 98, 13–19, Supplement 1.

Olsen, S. (1985) *Origins of the Domestic Dog: the Fossil Record*. Tucson, University of Arizona Press.

Ruusila, V. and Pesonen, M. (2004) Interspecific cooperation in humans (Homo sapiens) hunting: the benefits of a barking dog (Canis familiaris). *Annales Zoologici Fennici* 41, 545–549.

Sablin, M. V. and Khlopachev, G. A. (2002) The earliest Ice Age dogs: evidence from Eliseevichi I. *Current Anthropology* 43(5), 795–799.

Saetre, P., Lindberg, J., Leonard, J. A., Olsson, K., Pettersson, U., Ellegren, H., Bergstrom, T. F., Vila, C. and Jazin, E. (2004) From wild wolf to domestic dog: gene expression changes in the brain. *Molecular Brain Research* 126(2), 198–2006.

Savolainen, P., Zhang, Y. P., Luo, J., Lundberg, J. and Leitner, T. (2002) Genetic evidence for the east Asian origin of domestic dogs. *Science* 298(5598), 1610–1613.

Snyder, L. M. and Moore, E. A. (2006) (eds.) *Dogs and People in Social, Working, Economic or Symbolic Interaction*. Oxford, Oxbow Books.

Sundqvist, A. K., Bjornerfeldt, S., Leonard J. A., Hailer, F., Hedhammar, A., Ellegren, H. and Vila C. (2006) Unequal contribution of sexes in the origin of dog breeds. *Genetics* 172 (2), 1121–1128.

Tchernov, E. and Valla, F. F. (1997) Two new dogs and other Natufian dogs from the southern Levant. *Journal of Archaeological Science* 24(1), 65–95.

Terashima, H. (1983) Mota and other hunting activities of the Mbuti archers: A socio-ecological study of subsistence technology. *Kyoto University African Study Monographs* 3, 71–85.

Turnbull, C. M. (1965) The Mbuti Pygmies: An ethnographic survey. *Anthropological Papers of the American Museum of Natural History* 50(3), 139–282.

Turnbull, P. and Reed, C. (1974) The fauna from the terminal Pleistocene of Palegawra Cave, a Zarzian occupation site in northeastern Iraq. *Fieldiana Anthropology* 63(3).

Verardi, A., Lucchini, V. and Randi, E. (2006) Detecting introgressive hybridization between free-ranging domestic dogs and wild wolves (*Canis lupus*) by admixture linkage disequilibrium analysis. *Molecular Ecology* 15(10), 2845–2855.

Verginelli, F., Capelli, C., Coia, V., Musiani, M., Falchetti, M., Ottini, L., Palmirotta, R., Tagliacozzo, A., De Grossi Mazzorin, I. and Mariani-Costantini, R. (2005) Mitochondrial DNA from prehistoric canids highlight relationships between dogs and south-east European wolves. *Molecular Biology and Evolution* 22(12), 2541–2551.

Vila, C., Savolainen, P., Maldonado, J. E., Amorinim, I. R., Rice, J. E., Honeycutt, R. L., Crandall, K., Lundeberg, J. and Wayne, R. K. (2002) Multiple and ancient origins of the domestic dog. *Science* 276, 1687–1689.

White, F. (1983) *The Vegetation of Africa*. Paris, UNESCO.

Introduction and Methods

3. Past and present strategies for draught exploitation of cattle

Niels Johannsen

Draught exploitation of domestic cattle has been a major factor in the socio-economic development of many human societies. In the Old World, this technology can be traced back at least as far as the 4th millennium BC, while it was introduced much more recently in the New World. In general, in its attempts to investigate past exploitation strategies, zooarchaeology has relied on mortality patterns and possible indications of castration. It has sometimes been assumed that draught exploitation will manifest itself in faunal assemblages as a relatively high proportion of bones from old (castrated) males. However, this assumption is not necessarily correct in all contexts. Historical sources, ethnographic observations and literature on modern agro-economic and farming development reveal substantial variation in known draught exploitation strategies. This variation highlights to zooarchaeology the importance of supplementing mortality patterns with other lines of evidence, such as artefacts, iconography and osteomorphology.

Keywords: draught cattle, exploitation strategies, variation, zooarchaeological approaches

Introduction

The technology of harnessing cattle and exploiting their muscular power is one which has been of major socio-economic importance in many parts of the world. In many regions of the Eurasian continent, as well as in northern Africa, it has been practiced for millennia (Fansa and Burmeister 2004; Köninger *et al.* 2002; Pétrequin *et al.* 2006). In other areas, such as the Americas and large parts of Africa and Southeast Asia, the technology was introduced by European colonialists much later – in some places as recently as the 20th century AD (León *et al.* 2000; Starkey 1985; 1991; 1993).

Today cattle remain one of the most important sources of power in agriculture and transport around the world. Estimates indicate that more than 200 million cattle are utilized for traction on a world scale (Starkey 1991, 156). And while the technology is rapidly disappearing in some regions, it remains central or even becomes more widespread in other regions, particularly because of its economic sustainability (Hoffmann *et al.* 1989; Kalima 1994; Matthewman 1987).

The strategies by which the tractive power of cattle has been exploited, in the past as well as in recent times, vary substantially. This paper provides an introduction to and some examples of this variation in known strategies of exploitation. The examples given and general points made below draw on historical sources, ethnographic observations and literature on modern agro-economic and farming development. The paper is about practices in the employment of the most widespread domestic species of the genus *Bos*, *B. taurus* and *B. indicus*, though the use of other domestic bovines, such as water buffalo (*Bubalus bubalus*), shares many features (see, for instance, Hoffmann *et al.* 1989). The purpose of the discussion is to draw the attention of zooarchaeologists to variation in draught cattle utilization, and thereby to contribute towards informed investigations of the practice and role of this technology in past contexts. While the information provided does hold potential for specific analogies aimed at one or another delimited aspect of the technological practice, it is, above all, meant as a pool of background knowledge that zooarchaeologists and others studying past cattle husbandry may draw upon in their interpretations. The basic theoretical premise of this paper is that we will benefit as little from ignoring the information available on cultural variation in different types of human-animal relationships as we will from uncritically employing analogy (explicitly or implicitly).

Since draught exploitation of cattle is always part of an overall socioeconomic complex, the strategy adopted in a particular context depends on a large number of factors. This paper concentrates on variation found within a number of fundamental practice variables that are highly relevant to zooarchaeological interpretations of domestic cattle assemblages. Variation within the following variables will be discussed:

- The sex of the animals selected for work;
- The extent to which the animals are used for other, non-draught purposes;
- The ontogenetic age when training and exploitation begins;
- The methods that are used to train the animals, and the duration of this training;
- And, finally, the duration of the animal's work-life.

The treatment of these points below is primarily descriptive, though some discussion of the causal role of relatively simple functional considerations in strategic choices is offered. Other aspects of draught cattle technology – which are equally interesting to zooarchaeology – such as the impact of scales of socioeconomic organization or of variation in ideological preferences underlying technological practice, receive less attention here. Lastly, the effect of various draught exploitation strategies on zooarchaeological studies is discussed.

Sex and Purposes of Cattle Exploited for Traction

As illustrated below, the sex of cattle selected for work and the extent to which these animals are used for other, non-draught purposes are two tightly integrated elements of most draught exploitation strategies. Both female and male cattle have been employed in draught work since prehistoric times. Agricultural scenes decorating Egyptian tombs provide early evidence for both practices. For instance, a scene at the later 3rd millennium BC (Old Kingdom) Mastaba of Ḥetepḥerakhti, Sakkara, shows cows in front of a plough (Fig. 3.1), while a later 2nd millennium BC (New Kingdom) scene from the Tomb of Duauneḥeḥ, Thebes, shows male cattle being used for ploughing (Bartosiewicz et al. 1997, 25). Roman iconography, too, frequently depicts the use of draught cattle in agricultural activities. Ploughing scenes on Roman coins – a popular motif, probably related to the founding myth of Rome – seem to suggest that all such work was carried out by male cattle in this cultural context. Despite the focus of the iconography (reinforced, in the case of coins, by the medium, i.e. the repetitious use of a motif in minting), other evidence indicates that cows were used regularly for similar tasks in Roman agriculture. Columella (VI.24.3–5), writing in the 1st century AD, states that:

> There is no doubt that where there is great luxuriance of fodder, a calf can be reared from the same cow every year, but, where food is scarce, the cow must be used for breeding only every other year. This rule is particularly observed where cows are employed for work, in order that, firstly, the calves may have abundance of milk for the space of a year, and, secondly that a breeding cow may not have to bear the burden of work and pregnancy at the same time. When she has given birth to a calf, however good a mother she may be, if she is worn out by work, she denies the calf its due nourishment if her diet does not give her enough support.

In addition to providing evidence for Roman use of draught cows, this passage is interesting because it also illustrates that Roman farmers were faced with many of the same considerations as present users of draught cattle, trying to evaluate benefits and disadvantages of different exploitation strategies.

According to the historical sources and depictions available from the Roman period and onwards in Western Eurasia, it appears that the use of oxen (castrated male cattle) has been the solution most commonly chosen in this region (e.g. Columella VI.24; Culley 1807; Stokes 1853; Viires 1973). As hinted above, however, the employment of cows may not have been uncommon at all, and this raises the question whether the written and iconographic records come close to reflecting actual practice or present a distorted picture. The use of the stronger oxen may have been the practice that was regarded most highly and therefore the one given most attention in writing and iconography. Nonetheless, it seems that today, the employment of oxen (castrated males) is the preferred strategy in many parts of the world (e.g. Fig. 3.2).

Fig. 3.1 Cows used for ploughing in Old Kingdom Egypt. Section of a relief from the Mastaba of Ḥetepḥerakhti, Sakkara. After Bartosiewicz et al. 1997, 25.

However, the use of multi-purpose cows is equally a very common (if not dominant) strategy in some regions, such as certain African countries, and, especially, many parts of Asia (Matthewman 1987; Starkey 1993). Although cows are not as strong as oxen, they have the additional potential of producing calves and/or supplying milk, while still being available for work through much of the year, though draught work has a negative effect on fertility and milk production (Pearson *et al.* 1999). In other words, opting for female draught cattle implies certain functional advantages and respective disadvantages. The lesser strength of cows may limit their suitability for the heaviest tasks, such as logging or ploughing of very heavy soils. Furthermore, cows cannot (that is, should not) work during late pregnancy. But cows posses a greater purpose flexibility and are therefore especially suitable where the need for tractive power is more episodic, where ploughing and other tasks are relatively light, and where small farms with very limited resources need to maximize calorific efficiency as much as possible, for instance by getting tractive power, milk and the occasional calf from the same one or two animals. For example, farm size is an important factor in the overall exploitation pattern found in Bangladesh, where cows are most often used on small farms, while oxen are generally used on larger farms (Islam 1993, 124). Prioritizing sustainability through the use of multi-purpose cows is a solution of particular pertinence where suitable food for ruminants is scarce and the general level of pressure on resources high (Matthewman *et al.* 1990, 125–126). Together with the Roman evidence mentioned above, the situation in Bangladesh illustrates that different patterns of exploitation may make sense within different parts of the same overall cultural context.

The choice between female and male draught cattle is, in most cases, a choice between cows and oxen. This is, in turn, most often a choice between multi-purpose animals and single-purpose draught animals. Single-purpose oxen may be a good solution where there is plenty of draught work for much of the year and where heavy tasks are common. The term "single-purpose" is used here in the sense that the animal is utilized more or less exclusively for draught work (which of course encompasses many different activities) *until its death*, when another set of potential resources becomes available (meat, sinews, bones, hide *etc.*).

Male cattle selected as draught animals are, most often, castrated in order to enhance their tractability. The ontogenetic age at which castration is carried out varies substantially. The Carthaginian scholar of agriculture Mago recommended that castration is carried out when animals are "still young and tender". However, in cases where animals had already reached some age, he considered it better to castrate at the age of 2 years than at the age of 1 year (cited in Columella VI.24.5). British post-medieval sources recommend that castration should have already been carried out by the time the bull calf is 10 to 20 days old, and consider this as the best way to minimize the health risks associated with castration (Fitzherbert 1534, 58; Mortimer 1712, 162). Eastern Baltic sources from the 17th–19th centuries indicate that different traditions existed in different parts of this region. Some authors recommend castration between the age of a few weeks and a few months, while others emphasize the possibility of using a young bull in mating cows before castrating at the age of two or three years (Viires 1973, 440–441). Today, in many parts of the world the castration of potential draught animals appears to be delayed until the period before training commences, for instance in Southern India, where castration around the

Fig. 3.2 Peruvian farmer ploughing with a pair of oxen near the shores of Lake Titicaca, 2002. Photograph by the author.

age of 2 years is common (Panneerselvam and Kandasamy 1999, 17). One argument for postponing castration to the age of 2 years is to allow the natural level of male hormones to contribute to the development of a stronger neck (Kanu and Sankoh 1990, 138). In some recent African contexts, most animals selected for work are not castrated until they are 3 years or older (Barrett *et al.* 1982, 35).

There is some evidence for the employment of bulls, *i.e.* entire male cattle, in draught work in certain contexts. For instance, a Bronze Age rock carving from Litsleby in Sweden seems to portray mature bulls (recognizable by their erect phalluses) dragging an ard/scratch plough (Glob 1951, 26). Just like cows, such draught bulls would be able to fulfil a reproductive role as well. However, erect phalluses (of humans and animals alike) represent a very common motif in Scandinavian Bronze Age rock art, and it is questionable whether their massive iconographic presence is representative of daily life. Portraying fertile cattle may well have been much more important in the iconographic context than getting actual exploitation patterns right. But the possibility remains that bulls did in fact supply traction during this period, since this kind of practice is reported for some regions today, such as South and Central America (Starkey 1993, 68) and South Korea (Matthewman 1987, 216). In many cases, though, more detailed information on the use of bulls refers to the use of sub-adult individuals, which are more tractable than adults. One example of this practice is Bartosiewicz *et al.*'s (1997, 17) observation of a draught team consisting of a 6-year-old cow and her young, a bull aged less than 1 year, in front of a light wagon in the Buzău region of eastern Romania. Today, in most parts of the world, the draught employment of bulls which have reached full sexual maturity appears to be relatively rare (Lawrence and Pearson 2002; Starkey 1993).

To sum up, it is clear that both male and female cattle have been used for draught purposes in many *different* cultural contexts – both past and present. In addition to the general economic factors focused upon above, there are, of course, other factors that may decisively influence the chosen strategy. Culturally specific preferences for one solution or another, which are rooted in local ideology, often overshadow concerns about calorific efficiency. For example, Hindus for religious reasons generally avoid using cows for work (Lawrence and Pearson 2002, 107). In short, some of the important factors involved in selecting cattle for work are:

- The size and economic situation/wealth of the farm;
- The frequency of tasks;
- The nature of the tasks, and;
- Ideological preferences.

Ontogenetic Age When Exploitation Begins

If we turn now to the age at which the exploitation of individuals selected for draught work begins, we also find a fair amount of variation. Employment in regular tasks normally follows relatively soon after the training has been initiated, so whether one talks about the timing of training or exploitation, there is not a big difference in terms of ontogenetic age (we return to this point below).

The most common practice in both the past and present seems to be to commence training when the animals are 2 to 3 years old. This is, for example, the impression given by British historical sources (Mascall 1680, 199, 214; Stokes 1853, 12). Similarly, farmers near Lake Titicaca informed the present author that they would put oxen to work from around the age of 2½ years or "when the horns are long enough". While employing animals from the age of 2 to 3 years seems to be quite common, there are also many examples of exploitation commencing earlier, or even later. Cattle aged less than a year have been observed performing draught work (Bartosiewicz *et al.* 1997, 17). In Zimbabwe, the training of cattle under normal conditions begins when the animals are 1½ to 2½ years old (Neugebauer *et al.* 1989, 228). And in Tanzania animals raised under good conditions may be used from around 2 years, while animals raised under harsh conditions often do not start working until they are as much as 5 years old (Mgaya *et al.* 1994, 140). In 19th century Finland, the training of oxen normally commenced at 1½ to 2 years of age, while 3 years is reported as the norm in Estonia and Lithuania (Viires 1973, 441).

To sum up, the draught exploitation of cattle may commence within an ontogenetic spectrum of several years. As the evidence from Tanzania highlights, this is not only due to differing human preferences, but also to variation (according to environment, nutrition and breed) in the age at which maturity is reached by individual cattle.

Methods and Duration of Training

The training of draught cattle does not exhibit quite the same amount of variation as that encountered in the variables discussed so far. Basically, the difficulty of training draught cattle depends on whether or not the farmer has access to cattle which are already experienced in traction. It is indeed possible to train young cattle for traction without having access to other draught animals. This is done as a very gradual process beginning with holding the animals on a rope loop and leading them to walk or tying them up together if they are unmanageable. The next step would be to harness the animals and walk with them, also training verbal commands. Then the animals must be accustomed to pulling a load, before training with an actual plough or other implements can be initiated (Columella VI.1–2; Conroy 1999). Kanu and Sankoh (1990, 138) report from Sierra Leone that such training may take place over 3–4 weeks, while Djang-Fordjour *et al.* (2003, 3) suggest a duration of 2–3 weeks in Ghana.

The process summarized above represents the most difficult and the most time-consuming way of training draught cattle. So it is perhaps not surprising that the alternative method, which is to use an experienced animal as guide for the trainee, appears to be common in past and present contexts alike. Again, Columella (VI.2.6–11) explained this well in the 1st century AD. After having discussed how young cattle *can* be trained if no animals with draught-experience are present, he goes on to emphasize that:

> The method which we are prescribing should be followed only if no ox is available which has already done service; otherwise the system of training which we follow on our own farm is more expeditious and safer. For when we are accustoming the young bullock to the wagon or plough, we yoke with the untrained animal the strongest and at the same time quietest of the trained oxen, which both keeps it back if it rushes forward and makes it advance if it lags behind.

The method that Columella recommends here is known from a range of other cultural contexts, for instance the Baltic countries in the 19th century (Viires 1973, 441) and Serbia in the 1970s (Tony Legge pers. comm.). Just a few years ago, farmers near the shores of Lake Titicaca informed the author that this was the way they would normally train oxen. In most cases, animals trained by this method will start to contribute to the traction effort within a few days, after which the quality of their work will improve rapidly. So, by this method cattle are trained for a relatively short period of time before they are ready to perform regular work more or less satisfactorily. Naturally, the animals will continue to become more experienced for some time after they start working. But it is important to note that it is generally *not* a very slow and costly undertaking to make draught animals out of cattle. This point has implications for the last variable that we will take a look at.

Duration of Work-Life

Significant variation is found in the duration of draught cattle's work-life – that is, in the number of years that animals work before retiring (most often into the cooking pot). In some contexts, multi-purpose cows and oxen may continue to work until they are as much as 15 or in some cases even 20 years old, *e.g.* some of the Romanian oxen studied by Bartosiewicz *et al.* (1997). Culley (1807, 53) indicates that in 19th century England, cattle were mostly worked until 6 or sometimes 7 years old, after which they were fattened and slaughtered. Another Englishman, John Stokes, similarly reports that his practice was to sell oxen for fattening and butchering after 4 years of work (Stokes 1853, 9). In the Baltic countries, such practices were also prevalent in the 19th century. Here many oxen were sold around the age of 10, but it was also common to sell them for butchering at the age of only 6 or 7 years (Viires 1973, 443).

Perkins and Semali (1989, 298) report from Indonesia that most farmers retain cattle used for draught purposes for 2 to 4 years, and that female animals used for draught are generally retained for longer than oxen. In some cases – found both in Asian and South American contexts – farmers simply buy (or rent) a pair of animals at the beginning of the cultivation season, use and fatten them, and then sell them by the end of the season (Lawrence and Pearson 2002, 105; Sims *et al.* 2002, 2). In such cases, micro-economic fluctuations are likely to determine the number of working seasons that a given draught individual will work before being (sold and) slaughtered. Indeed, small-scale subsistence farmers in the Titicaca region of Bolivia and Peru have emphasized to the author the relevance of such fluctuations in the situation of the individual farm in determining the duration of a draught animal's work-life. An ox might be slaughtered after having worked for a year or two, or it might work until it is very old. Some of the factors involved are how well the animal performs during work and whether there are people who are interested in buying animals at a good price. But, above all, the duration of the animal's work-life depends on how well the family is doing in general – in other words which resources are available to the family – for instance whether they have other animals that can be eaten or provide tractive power. Similar observations apply to the management of draught cattle in Upper Volta (Barrett *et al.* 1982) and Ethiopia (McCann 1984) and to the management of buffalos in Nepal (Lawrence and Pearson 2002, 103).

Concluding Remarks

There are a number of points we can draw out of the discussion above, which are particularly important for zooarchaeology:

- Firstly, we can note that both female and male cattle are potential draught animals from quite an early age.
- Secondly, in societies which are accustomed to this technology, draught cattle are generally trained relatively easily.
- Thirdly, individual draught cattle may be used for a few years or until they are too old to work.
- Finally, in many cases strategies are not all that rigid – they are frequently customized for changes in the specific economic situation that the farmer is facing.

What are the zooarchaeological implications of these points? The main implication is perhaps, simply put, that we have to keep in mind that when we examine draught exploitation of cattle in the zooarchaeological record, we are examining a highly variable phenomenon. This is particularly important to remember in the analysis of mortality patterns, which is the most common – and in many ways most essential – zooarchaeological tool for studying exploitation strategies. We cannot have a set of fixed expectations with regard to what draught exploitation of cattle might look like in a mortality pattern, and in many cases this type of exploitation may be very difficult to discern among all the factors contributing. What this highlights is the importance of supplementing mortality patterns with as many other lines of evidence as possible. This is, of course, a fairly banal point for zooarchaeology in general. With regard to cattle traction, such lines of evidence may include written sources and iconography, as exemplified above; they may include artefacts and structures which are part of draught animal technologies (*e.g.* Fansa and Burmeister 2004; Köninger *et al.* 2002; Pétrequin *et al.* 2006); and they may include investigations of changes in osteomorphology (*e.g.* Bartosiewicz *et al.* 1997; Johannsen 2005; 2006). Of course, all of these other lines of evidence present their own methodological challenges and difficulties, but that is a different story.

Acknowledgements

I would like to thank the participants in the *Ethnozooarchaeology* session at the 2006 ICAZ International Conference in Mexico City for inspiring discussions, and László Bartosiewicz for providing the image used in Fig. 3.1. Most of the British historical sources drawn upon in this paper were brought to my attention by an article by Simon Davis (2002).

References

Barrett, V., Lassiter, G., Wilcock, D., Baker, D. and Crawford, E. (1982) *Animal Traction in Eastern Upper Volta: A Technical, Economic and Institutional Analysis*. MSU International Development Papers 4. East Lansing, Department of Agricultural Economics, Michigan State University.

Bartosiewicz, L., Van Neer, W. and Lentacker, A. (1997) *Draught cattle: their osteological identification and history*. Annales Sciences Zoologiques 281. Tervuren, Musée Royal de l'Afrique Centrale.

Columella, L. J. M. (1941–1955) *De Re Rustica*. Translation by H. B. Ash, E. S. Forster and E. H. Heffner. London, Heinemann.

Conroy, D. (1999) *Oxen: A Teamsters Guide*. Gainesboro, Rural Heritage.

Culley, G. (1807) *Observations on Live Stock, Containing Hints for Choosing and Improving the Best Breeds of the Most Useful Kinds of Domestic Animals* (4th edition). London, Wilkie and Robinson.

Davis, S. J. M. (2002) British Agriculture: Texts for the Zoo-Archaeologist. *Environmental Archaeology* 7, 47–60.

Djang-Fordjour, K. T., Asare-Mantey, G., Bediako, J. A. and Otchere, E. O. (2003) The uses and management practices for draught animals in the Northern Zone of Ghana. *Draught Animal News* 39, 2–5.

Fansa, M. and Burmeister, S. (eds.) (2004) *Rad und Wagen: Der Ursprung einer Innovation. Wagen in Vorderen Orient und Europa*. Archäologische Mitteilungen aus Nordwestdeutschland 40. Mainz am Rhein, Philipp von Zabern.

Fitzherbert, A. (1534) *The Boke of Hvsbandry*. London, Thomas Berthelet (edition edited by W. W. Skeat, 1882, London, Trubner).

Glob, P. V. (1951) *Ard og Plov i Nordens Oldtid* (Ard and plough in prehistoric Scandinavia). Jutland Archaeological Society Publications 1. Aarhus, Aarhus University Press.

Hoffmann, D., Nari, J. and Petheram, R. J. (eds.) (1989) *Draught Animals in Rural Development*. ACIAR Proceedings No. 27. Canberra, Australian Centre for International Agricultural Research.

Islam, A. W. M. S. (1993) Draught Animal Power in Bangladesh. In W. J. Pryor (ed.) *Draught Animal Power in the Asian-Australasian Region*. ACIAR Proceedings No. 46, 123–130. Canberra, Australian Centre for International Agricultural Research.

Johannsen, N. N. (2005) Palaeopathology and Neolithic cattle traction: methodological issues and archaeological perspectives. In J. Davies, M. Fabiš, I. Mainland, M. Richards and R. Thomas (eds.) *Health and Diet in Past Animal Populations*, 39–51. Oxford, Oxbow.

Johannsen, N. N. (2006) Draught cattle and the South Scandinavian economies of the 4th millennium BC. *Environmental Archaeology* 11, 33–46.

Kalima, C. (1994) Animal traction technology for logging in Zambia. In P. H. Starkey, E. Mwenya and J. Stares (eds.) *Improving animal traction technology*, 445–447. Wageningen, Technical Centre for Agricultural and Rural Cooperation.

Kanu, B. H. and Sankoh, F. A.-R. (1990) Management of Work Oxen in Sierra Leone and its Implications for Research. In P. R. Lawrence, K. Lawrence, J. T. Dijkman and P. H. Starkey (eds.) *Research for Development of Animal Traction in West Africa*, 137–139. Addis Ababa, The International Livestock Centre for Africa.

Köninger, J., Mainberger, M., Schlichtherle, H. and Vosteen, M. (eds.) (2002) *Schleife, Schlitten, Rad und Wagen*. Hemmenhofener Skripte 3. Gaienhofen-Hemmenhofen, Landesdenkmalamt Baden-Württemberg.

Lawrence, P. R. and Pearson, R. A. (2002) Use of draught animal power on small mixed farms in Asia. *Agricultural Systems* 71, 99–110.

León, A. C., Saldaña, T. M. and Miranda, C. R. (2000) La tracción animal en México. *Draught Animal News* 33, 5–9.

Mascall, L. (1680) *The Countreyman's Jewel: or, the Government of Cattel*. London, William Thackery.

Matthewman, R. W. (1987) Role and Potential of Draught Cows in Tropical Farming Systems: A Review. *Tropical Animal Health and Production* 19, 215–222.

Matthewman, R. W., Dijkman, J. T. and Zerbini, E. (1990) The Management and Husbandry of Male and Female Draught Animals: Research Achievement and Needs. In P. R. Lawrence, K. Lawrence, J. T. Dijkman and P. H. Starkey (eds.) *Research for Development of Animal Traction in West Africa*, 125–136. Addis Ababa, The International Livestock Centre for Africa.

McCann, J. (1984) *Plows, Oxen, and Household Managers: a Reconsideration of the Land Paradigm and the Production Equation in Northeast Ethiopia*. African Studies Center Working Papers No. 95. Boston, Boston University.

Mgaya, G. J. M., Simalenga, T. E. and Hatibu, N. (1994) Care and management of work oxen in Tanzania: initial survey results. In P. H. Starkey, E. Mwenya and J. Stares (eds.) *Improving animal traction technology*, 139–143. Wageningen, Technical Centre for Agricultural and Rural Cooperation.

Mortimer, J. (1712) *The Whole Art of Husbandry; or, the Way of Managing and Improving of Land* (3rd edition). London, H. and G. Mortlock.

Neugebauer, J., Chikwanda, B. and Magumise, J. (1989) Training and Handling of Oxen for Work in Zimbabwe. In D. Hoffmann, J. Nari and R. J. Petheram (eds.) *Draught Animals in Rural Development*. ACIAR Proceedings No. 27, 226–230. Canberra, Australian Centre for International Agricultural Research.

Panneerselvam, S. and Kandasamy, N. (1999) Physical characters and load hauling capacity of Kangayam bullocks of South India. *Draught Animal News* 31, 17–20.

Pearson, R. A., Zerbini, E. and Lawrence, P. R. (1999) Recent advances in research on draught ruminants. *Animal Science* 68, 1–17.

Perkins, J. and Semali, A. (1989) Economic Aspects of Draught Animal Management in Subang, Indonesia. In D. Hoffmann,

J. Nari and R. J. Petheram (eds.) *Draught Animals in Rural Development*. ACIAR Proceedings No. 27, 294–299. Canberra, Australian Centre for International Agricultural Research.

Pétrequin, P., Arbogast, R.-M., Pétrequin, A.-M., van Willigen, S. and Bailly, M. (eds.) (2006) *Premiers chariots, premiers araires. La diffusion de la traction animale en Europe pendant les IVe et IIIe millénaires avant notre ère*. CRA Monographies 29. Paris, CNRS Éditions.

Sims, B. G., Zambrana, L. and Dijkman, J. (2002) Improved management and use of draft animals in the Andean hill farming systems of Bolivia. *Draught Animal News* 36, 2–4.

Starkey, P. H. (1985) Animal traction research and extension in Africa: an overview. In S. V. Poats, J. Lichte, J. Oxley, S. L. Russo and P. H. Starkey (eds.) *Animal traction in a farming systems perspective*. Farming Systems Support Project Network Report No. 1. Gainesville, Institute of Food and Agricultural Sciences, University of Florida.

Starkey, P. H. (1991) Draught Cattle World Resources, Systems of Utilization and Potential for Improvement. In C. G. Hickman (ed.) *Cattle Genetic Resources*. World Animal Science B7, 153–200. Amsterdam, Elsevier.

Starkey, P. H. (1993) A world view of animal traction highlighting some key issues in eastern and southern Africa. In P. H. Starkey, E. Mwenya and J. Stares (eds.) *Improving animal traction technology*, 66–81. Wageningen, Technical Centre for Agricultural and Rural Cooperation.

Stokes, J. (1853) *The Ox as a Beast of Draught in Place of the Horse, Recommended for all Purposes of Agriculture, After a Satisfactory Experience of Thirty Years' Employment of Ox Labor*. London, Pelham Richardson.

Viires, A. (1973) Draught Oxen and Horses in the Baltic Countries. In A. Fenton, J. Podolák and H. Rasmussen (eds.) *Land Transport in Europe*. Studies of Folklife 4, 428–56. Copenhagen, The National Museum of Denmark.

4. Animal dung: Rich ethnographic records, poor archaeozoological evidence

Marta Moreno-García and Carlos M. Pimenta

Ethnographic work among traditional rural societies shows that animal dung is not a waste resource. Not only does its collection form part of the daily list of tasks, but also many other activities, important for the local economy, depend on it. The two main uses described worldwide are: fertilizer for agricultural fields and fuel for domestic fires. A review of the literature illustrates that the interest in recognising the occurrence of dung and its uses in the archaeological record is relatively recent. It is remarkable that, despite it being a product derived from animals, the use of dung has rarely been discussed by archaeozoologists. In fact, most of the research has been undertaken through botanical, palynological, sedimentological and micromorphological analyses. Ethnographic observations made among agro-pastoralist communities in northern Morocco demonstrate that direct evidence for the exploitation of this animal product is actually very poor. Hence, it is concluded that assessing the role of dung through archaeozoological analysis will be possible only if the study of faunal remains is made from a broad perspective that tries to recognise the relationship of husbandry techniques with environmental, social and economic circumstances.

Keywords: animal dung, ethnographic data, northern Morocco, archaeozoological evidence, secondary products

"While we tend to think of livestock mainly as a source of meat and milk, in practice they produce more dung than anything else."

(Sillar 2000, 46)

Introduction

In the last few decades, ethnographic research intended to improve our understanding and interpretation of archaeological data has shown that in regions where mechanization of agriculture and the use of chemical fertilizers are uncommon, dung plays a major role in peoples' daily life (Lewthwaite 1981; Miller 1984; Ertug-Yaras 1996; Anderson and Ertug-Yaras 1998; Sillar 2000; Zapata Peña *et al.* 2003). Livestock transform plant material in the environment into a resource that provides "the essential organic component to bind the soil matrix, reduce erosion and fasten any available nutrients so that they are not leached out of the soil too rapidly, while at the same time the dung gradually releases its own vital plant nutrients as it decays" (Sillar 2000, 52). Moreover, in arid or deforested areas where there is a scarcity of wood, animal dung is burnt and constitutes the main kind of fuel (Miller and Smart 1984; Anderson and Ertug-Yaras 1998; Sillar 2000). The close relationship between dung and traditional agriculture has prompted archaeobotanists to address these issues (Charles 1998) and particularly its use as fuel through the identification of charred plant remains in archaeological sites (Miller and Smart 1984; Wilson 1984; Anderson and Ertug-Yaras 1998; Charles 1998; Charles and Bogaard 2005). They have also shown an interest in distinguishing between crops grown for human food and those grown as fodder for animals (Jones 1998). In addition, palynological analyses of fossil dung have successfully documented past vegetation and environmental changes (Scott 1987; Carrión *et al.* 2000; di Lernia 2001). Sedimentological and micromorphological techniques such as the production of petrographic thin sections have been applied to determine the content of animal dung (Courty *et al.* 1991; Brochier *et al.* 1992; Akeret and Rentzel 2001; Korstanje 2005), to identify the spherulites present in ashy cave sediments derived from penned herbivores (Canti 1997; Canti 1998) and to provide evidence of how space was used in the past (Shahack-Gross *et al.* 2005; Karkanas 2006).

In contrast, production, procurement and use of dung are issues that tend to be overlooked in archaeozoological studies concerned with ancient livestock management. Other animal products such as meat, milk, wool, marrow, fat, bone and even draught power have earned most of the attention (Halstead 1998; de Cupere *et al.* 2000; Copley

Fig. 4.1 Location of the area of research in northern Morocco.

et al. 2003; Luik et al. 2003; Davies et al. 2005; Mulville and Outram 2005). Only relatively recently has the plant material present in dung been studied to investigate animal diet and its relationship to strategies of livestock exploitation (Charles 1998; Charles et al. 1998).

In general, the presence of suitable dung-producing animals in a faunal assemblage induces the archaeozoologist to assume the occurrence of this material, but its economic importance is seldom assessed. The reason for this probably lies in the difficulty of recognising the use of dung in the archaeozoological record. Variables such as the age at death or sex ratios employed to identify if herds were mainly exploited for their milk, wool or meat do not reflect dung production and utilisation.

Ethnozoological research provides the means to understand and evaluate the role of this animal product among traditional agro-pastoralist communities. This paper describes a case study among self-sufficient rural communities in the Jebala region, located at the westernmost part of the Rif in northern Morocco (Fig. 4.1), with two aims: i) to show the multiple uses made of dung in these peoples' daily life, and ii) to identify which aspects related to its exploitation one might expect to be detectable in the archaeozoological record. Fieldwork was undertaken between 1999 and 2001 and was part of an ethnoarchaeological project which focused on agrarian systems (Ibáñez Estévez et al. 1999), domestic and wild animal resources (Moreno-García 2004a, b; Pimenta 2004), ceramic (González Urquijo et al. 2001; Ibáñez et al. 2001b) and leather technology (Ibáñez et al. 2001a).

Animal Dung: A Highly-Estimated Commodity in the Jebala Mountains, Northern Morocco

The Jebala region (Morocco) is a mountainous area with altitudes above 2,000m and a Mediterranean climate. Precipitation varies with the topography and in the highest areas exceeds 2,000mm per year. Although deforestation is occurring gradually, the landscape is still dominated by woodlands of lentisk (*Pistacia lentiscus*), oaks (*Quercus suber, Q. ilex, Q. coccifera/rotundifolia*) and strawberry tree (*Arbutus unedo*). Small villages (*duar*) of up to 20 families are located on the mountain slopes or in the valley bottoms. Most are difficult to reach because there are no proper roads but only sand tracks. The observations discussed here were recorded in small villages of ancient qabila *Gzaua* of northern Morocco (province of Chefchauen) (Fig. 4.1) through interviews with several families and the help of translators. Each of these families lives on the small agricultural fields they own and off their livestock, mainly goats and cattle. Economically they are essentially self-sufficient and practise little external trade.

Cattle and donkeys are employed for traction in agricultural tasks such as ploughing and threshing. These activities are exclusively performed by men. Cattle are considered a banking resource (Halstead 1993) and are valued for their 'secondary products', but not all families own them. They are stabled during the autumn and winter, and are fed with fresh fodder brought in daily by the women. Goats are more common than cattle although their numbers per family tend to vary. Since it is the women in the house, and particularly the teenage girls, that shepherd the flocks, the families with unmarried daughters can afford to keep a larger number of heads than those whose daughters have married and left home.

Among the inhabitants of the Jebala, meat consumption of their own livestock is occasional. It is usually related to special social or religious events, which does not mean that they do not turn to other available resources such as wild animals (Moreno-García 2004a). Goat flocks are constituted by a range of different-aged animals, but sub-adult and adult females tend to be more common. One-year old males are usually sold in the market or are sacrificed at family events. Dairying is much more common, although it is a seasonal activity, practised on a small scale and for family use due to the small number of animals in each flock (Moreno-García 2004b). Hence, the only product they have to deal with on a daily basis is dung.

Dung Collection and Preservation

In all villages goat dung is more common than that of cattle since goat numbers are higher. Goats are kept during the night in a covered stable with a patio attached to the house. In spring and summer they are taken out in the morning and in the afternoon, avoiding the hottest hours of the day, to the sloped more forested surroundings of the village to feed freely. Thus, the stable is swept when they leave and droppings are collected by the women twice a day. In winter and autumn this task is performed only once in the morning, because the animals are not returned at midday. Droppings are deposited a few meters from the house, in the open sun, forming a small pile that is surrounded by

Fig. 4.2 Accumulation of goat droppings in the vicinity of a house. Kalaah (northern Morocco).

Fig. 4.4 Cowpats drying on a rooftop. Briet (northern Morocco).

Fig. 4.3 Stockpile of goat dung spread along a slope at the back of a house. Kalaah (northern Morocco).

a low wall of stones and interwoven branches to prevent animals and people from stepping on it (Fig. 4.2). In Kalaah they showed us a large stockpile of goat dung situated on a slope at the back of the house. Droppings were thrown from the top of the slope and rolled down to form a dense accumulation at the bottom (Fig. 4.3). Charcoal and ash from domestic hearths and baking ovens were also disposed here, contributing towards the decomposition of the manure and adding nutrients.

Cow dung is also collected from the stables, although not on a daily basis. It is usually mixed with the straw that serves as bedding for the cattle. The mixture is stored in heaps or in swallow pits, protected by branches and twigs, outside each family home. Cow pats are preferably used fresh, so not all families spend time making dry dung cakes. It is generally in late spring or during the summer, before the pottery season begins, that they can be seen drying on the rooftop (Fig. 4.4). Since woodland areas are still abundant, wood is the main source of domestic fuel.

It is interesting that each family has its own store of decomposing goat and/or cattle dung for use as fertilizer. However, when it comes to the availability of fresh cattle dung to be used as such (see below), anybody can pick up the pats of the cattle that are feeding freely in the field.

Dung as Manure, Fertilizer of the Local Crop Fields

No chemical fertilizers are used by these rural communities. The fields and orchards they cultivate close to home are small. After it has decomposed, the manure obtained from their own livestock is transported to the fields in baskets on the backs of women or donkeys. Usually cattle and goat dung are mixed before being spread as manure. The reason is that goat dung applied on its own is so energetic that it may burn fruit trees unless there is lot of rain. The villagers of the Jebala recognise the different properties of each animal's dung as recorded in the 11th–12th centuries Hispano-Arabic Treatises of Agriculture such as those of Ibn al-'Awwâm (Banqueri 1802), Ibn Wafid (Millás Vallicrosa 1943) and Ibn Bassal (Millás Vallicrosa 1948; Millás Vallicrosa and Aziman 1955). For instance, Ibn Wafid mentions that the best livestock manure derives from horses, mules and donkeys, followed by that of sheep and goats, and lastly that of cattle. He recommends not using that derived from pigs because it will kill all plants. Ibn Bassal also notes that it is important that sheep manure is matured. Otherwise, when applied 'half done' there will

Fig. 4.5 Burning dry cowpats mixed with straw. The smoky mixture is introduced in a special vessel (on the left) that is used to drive away bees from the beehive. Budarna (Ouezzane region, Morocco).

Fig. 4.6 Firing of pottery using dry cow dung cakes as fuel. Ain Kob (northern Morocco).

be a proliferation of weeds because their tiny seeds are not altered in the sheep digestive system and can germinate.

We were also told that donkey dung is most appropriate for olive trees, and those families that have pigeons collect their dung and mix it with salt and lime for tanning skins. Even chicken dung is recommended for sick trees.

Dung as Fuel

In the Jebala region, the use of dung as fuel is restricted to two activities undertaken only by a small part of the population: beekeeping (a male activity; Pimenta 2004) and the firing of pottery (always carried out by women; González Urquijo *et al.* 2001). A general practice in the world of beekeeping is smoking the entrance of the hive before opening the top cover. Smoke calms bees; it initiates a feeding response in anticipation of possible hive abandonment due to fire, and also masks the alarm pheromone secreted by the guard bees waiting at the entrance, so the rest of the colony is not prone to sting (Visscher *et al.* 1995). In the Jebala, the fuel used consists of partially disaggregated cakes of dry cow dung, mixed with straw (Fig. 4.5). This mixture is ignited and placed inside a smoker – a special vessel called *nafuja* that has two mouths, one completely opened and the other closed but pierced. The smoker is then taken to the beehive which is set lying on the floor. The opened mouth is put in at its entrance while a man blows the smoke inside from the pierced end. Once the bees have been driven away they collect their honey.

Pottery is hand-made and fired in domestic hearths or open fires using dry and fresh cow dung cakes as fuel. The whole process has been described in detail by our colleagues (Zapata Peña *et al.* 2003) and small variations between different potters have been recorded in the area. For the purpose of this paper we only present here some general notes. The open fire is located some meters away from the house, in a small rounded depression covered with wood splinters and surrounded by a wood belt supported by a circle of stones. The containers to be fired are placed inside and are covered both with dry and fresh cowpats. Three openings are left at the base of the structure through which the fire is lit with the help of branches, and afterward they are closed with dry dung cakes (Fig. 4.6). Fresh dung is added during baking to cover any open area with flames. The process takes several hours and the fire is usually left overnight to cool down. Next morning, vessels are removed and any charcoal produced is collected, stored and re-used for cooking in the domestic hearths.

The reasons for using dung as fuel only in these two 'professional' activities are twofold. First, it burns more slowly than wood and provides a higher concentration of heat and slow cooling (Sillar 2000; Zapata Peña *et al.* 2003). Second, the amount of dung consumed in apiculture and firing of pottery is small and barely affects its availability for other more general uses (*e.g.* manure for the fields or tempering material for walls and floors of the houses).

Dung as Tempering/Insulating or Construction Material

After manure, the most widespread use of cattle dung is as a material for tempering and insulating. All floors of the house and adjacent areas, such as open-air stables or threshing floors are coated with a mixture of fresh cowpats diluted in water. Mixed with clay, they are used to insulate the walls of the houses. This is a seasonal activity carried out during the summer to facilitate drying (Fig. 4.7). In Italy, Spain, and Portugal the ethnographic literature indicates that in these countries, threshing floors were also tempered with cattle dung in recent times (Lopes Dias 1970; Viegas Guerreiro 1982; Viegas Guerreiro *et al.* 1982). In the case of Portugal, Fontes (2005) mentions that in the northern region of Trás-os-Montes threshing floors were communal property, thus the whole population was mobilized to collect dung for this purpose. Three months before their tempering, men, women and even children "*iam à merda*", literally translated "they went to the shit". Fresh cowpats were collected several times a day, preferably at the very moment they were produced by the animal. It was

Fig. 4.7 Coating of the walls with fresh cow dung mixed with clay. (Ouezzane region, Morocco).

important that they were as clean as possible and not mixed with soil or straw. They were accumulated to form a pile that grew week by week until the day they were moved to the threshing floor that needed tempering.

Fresh cattle dung is also one of the materials employed in the construction of the two types of beehives in the Jebala region. Due to its plasticity, it is used to seal the entrance and gaps of the beehives fashioned from cork, and to plaster those made of interwoven twigs (Fig. 4.8). Both are manufactured by men only. Similar practices have been noted in Northern Ethiopia, Turkey and India where storage pits or baskets used as grain containers are wrapped with cattle dung (Rawat *et al.* 2000). In contrast, goat droppings do not appear to be used for any of those activities. Neither did we record the observation Jean-Denis Vigne (pers. comm.) made in northern Khorasan (Iran), an arid zone at the edge of the desert where trees and large stones are absent. There, the walls of enclosures and sheep stalls were built with 'dung bricks', blocks of dry dung and mud scraped from the floors of sheep pens (Fig. 4.9).

Dung Used for Making Containers

During the project we recorded a custom that is about to be abandoned by the Jebala potters: the use of cow dung on its own or mixed with clay to make containers (Fig. 4.10). Again, this was an activity carried out by women during the summer to facilitate drying. Two main types of containers were made: the *tonna*, a high wall cylindrical vessel that could reach 2m in height, used for storing dried foods (cereals, legumes, salt, *etc.*), and the *tabtoba*, a small wide open bowl used for feeding animals and for taking grain to the quern. The addition of cow dung has the advantage of making these vessels lighter than those made of clay alone (Ibáñez *et al.* 2001b). Fresh cow dung was taken directly from the stable or the storage pits near the houses. The production of a single container was a slow process that could take as many as 15 days since after joining and smoothing several coils together they had to be left to dry. When dung was too soft some clay would be added to increase the consistency of the paste.

Fig. 4.8 Beehives made of interwoven twigs and plastered with fresh cow dung. Budarna (Ouezzane region, Morocco).

Archaeozoological Assessment of the Ethnographic Data

The ethnographic observations described above show that dung can be an animal resource collected regularly and widely used by self-sufficient agro-pastoralist communities. It is certainly not a wasted resource. In this sense, ethnozoology provides a potential framework for analysing the exploitation of livestock as dung producers from an archaeozoological perspective. It seems to us that data related to issues such as species occurrence and frequency or husbandry strategies may be explored by the archaeozoologist to detect and assess the importance of this animal product in the archaeozoological record.

Frequency of Species

In any household in these rural communities of the Jebala region in northern Morocco, the dominant species in terms of number of heads is goat, followed by cattle and donkeys. Consequently, one could assume that the most abundant dung available would be the most widespread used, in this case that of goat. However, our ethnographic observations indicate that cattle dung is employed most extensively in a variety of tasks. This means that there is not necessarily

Fig. 4.10 Woman making containers out of cattle dung. Harrakah (northern Morocco).

Fig. 4.9 Walls fashioned from sheep dung 'bricks', resulting from scraped sheep pen floors. Khorasan, northern Iran. Courtesy of Jean-Denis Vigne.

a direct correspondence between the most abundant species and the dung used. In addition, the ethnographic observations indicate that not all dung is the same. Dung seems to differ according to which species of animal it is derived from, and also it has different uses in its different states – fresh, dried or rotten. For instance, the plasticity of cattle dung is undoubtedly a characteristic that enhances its multiple uses in opposition to that of any other kind of dung. Thus, in the archaeozoological record frequency of species may perhaps be a good indicator of the most common dung available, but it could be misleading in the identification of the most used or valued, especially in those sites where different livestock species were exploited.

Husbandry Strategies

Livestock management appears to be an important issue to consider when investigating dung production and use in the past. Domestic animals may be kept free-range or stabled. A particular human community may practise either husbandry strategy or a combination of both depending on several factors: the size of the herds/flocks, season of the year, availability of grazing land, and labour available. Thus, in the Jebala, each family that owns one or several cattle keep them penned and stall-fed with dry fodder during the winter, whereas in spring and summer they are taken to graze freely in the fields for the whole day until dusk, when they are returned to the stables. Goat flocks graze freely most of the year, being occasionally supplied with fodder in winter. They are also kept indoors during the night. Stalling conditions vary according to season, because during summer the animals may spend the night on the patio outside the stables. In both circumstances, penning next to the house at night results not only in daily and easy collection of dung by members of the family, but also facilitates its preservation in the surrounding area. Access to dung appears to be an essential issue to take into account when assessing its economic value for self-sufficient societies. Furthermore, it is important to understand that dung procurement for anything else other than fertilizer becomes a time consuming activity and requires some effort and labour. In this context, Sillar (2000, 49) notes that llamas and alpacas grazing in the higher areas of the Andean *altiplano* defecate in communal piles, often near a water source, making the collection of their dung very easy for families that make special trips there at the end of the dry season. Since this is a treeless region where wood is scarce, fresh or dry llama droppings are highly valued as fuel for domestic use and for firing traditional pottery.

Preference for one or another use of dung is also related to economic and environmental factors, as exemplified, for instance, by transhumant livestock in the Mediterranean. It is well known that in medieval Spain landowners of large extensive uncultivated fields in the plains welcomed transhumant sheep flocks to feed on them during the winter in exchange for the manure they produced, and for the same reason, in the summer grazing areas, they were allowed to enter the stubble fields together with the stationary flocks (Moreno-García 1999). Stubble represented a considerable supply of high nutritive value to be consumed by the animals after the summer pastures had been depleted.

In conclusion, traditional agriculture is absolutely dependent on animal manure. Its use as fertilizer was probably the most comprehensive. Notwithstanding, among self-sufficient economies there seem to be other uses of dung facilitated by the livestock management techniques employed, which could be explored in the archaeozoological record if the faunal analyst takes into account other variables such as environmental, economic and social conditions.

The Archaeozoological 'Invisibility' of Dung

Dung is a continuously available animal product for those communities engaged in livestock husbandry. And unlike milk, wool or traction power, its production is not dependent on the sex or the age of the animal. Male, female, castrate, juvenile, adult or mature individuals of any species produce dung daily. Therefore, the data obtained from vertebrate faunal remains regarding species identification and frequency, sex ratios and age at death alone do not provide information that could be directly interpreted as a reflection of dung exploitation in the past. Palaeoentomology is the only area within archaeozoology that has succesfully provided evidence of dung through the identification of invertebrates (*e.g.* mites, parasites, insects and so on) associated with it (Kenward and Hall 1997; Hall and Kenward 1998). In any case, due to the very nature of this substance, it is very difficult to assess its economic importance in antiquity through the archaeological record. Its invisibility contributes towards its being largely ignored in the archaeozoological literature.

Our ethnographic observations in the Jebala region of northern Morocco show that it is part of the self-sufficient nature of rural communities to produce for the local economy. People are dependent on their own production. Hence, manure is essential for sustaining the family's cultivated fields and orchards. No commercial or trading profits are involved in its production and use. Cattle cowpats are favoured for a variety of uses due to their plasticity and insulating properties. Men and women use them in activities that are performed seasonally. A combination of free-range and stalled husbandry of the small number of animals owned by each family represents an optimisation of resources. Not only is there easy access and control of the dung produced, but also little effort and manual labour is required to collect and manipulate it.

As in many other areas of archaeology, ethnographic studies have raised the visibility and the potential value of an animal product such as dung, which disappears with little trace from the archaeological record. The results presented here show that assessing its role through archaeozoological analysis will be possible only if the study of faunal remains is made from a broad perspective and tries to recognise the relationship of husbandry techniques with environmental, social and economic circumstances in a particular geographical area and at a specific chronological period.

It is also evident that this ethnoarchaeozoological approach suggests the need for more work on the subject in order to answer questions that we did not attempt to explore here. In this respect, collection of data regarding the amount of dung produced by each species of domesticated livestock in a particular period, how much of it is used in different activities and variations in dung quality/properties according to season of the year, sex and age of the animal present interesting avenues of investigation. Certainly, they will contribute towards enhancing the archaeozoological visibility of this animal product.

Acknowledgements

This work was part of the project "*Las primeras sociedades campesinas. El aporte de la etnoarqueología en Marruecos*" supported by the Spanish Fundación Marcelino Botín. We are very grateful to the Fundação Calouste Gulbenkian in Lisbon (Portugal) that subsidised one of the authors (MM-G) to participate in the 10th ICAZ Conference in México City. Also we would like to thank Umberto Albarella for inviting us to the Ethnoarchaeozoology session and for editorial comments. Jean-Denis Vigne is acknowledged for sharing with us unpublished information on his work in Iran and supplying Fig. 4.10. Thanks also to Naomi F. Miller and Bill Sillar for kindly sending copies of their work. Finally, we are grateful to Simon Davis for correcting the English and commenting on an earlier draft of this paper as well as to an anonymous referee.

References

Akeret, Ö. and Rentzel, P. (2001) Micromorphology and plant macrofossil analysis of cattle dung from the Neolithic Lake Shore settlement of Arbon Bleiche 3. *Geoarchaeology* 16, 687–700.

Anderson, S. and Ertug-Yaras, F. (1998) Fuel fodder and faeces: an ethnographic and botanical study of dung fuel use in Central Anatolia. *Environmental Archaeology* 1, 99–109.

Banqueri, J. A. (ed.) (1802) *Kitab al-filaha. Libro de agricultura de Ibn al-'Awwâm*. Madrid.

Brochier, J. E., Villa, P., Giacomarra, M. and Tagliacozzo, A. (1992) Shepherds and sediments: geo-ethnoarchaeology of pastoral sites. *Journal of Anthropological Archaeology* 11, 47–102.

Canti, M. G. (1997) An investigation of microscopic calcareous spherulites from herbivore dungs. *Journal of Archaeological Science* 24, 219–231.

Canti, M. G. (1998) The micromorphological identification of faecal spherulites from archaeological and modern materials. *Journal of Archaeological Science* 25, 435–444.

Carrión, J. S., Scott, L., Huffman, T. and Dreyer, C. (2000) Pollen analysis of Iron Age cow dung in southern Africa. *Vegetation History and Archaeobotany* 9, 239–249.

Charles, M. (1998) Fodder from dung: the recognition and interpretation of dung-derived plant material from archaeological sites. *Environmental Archaeology* 1, 111–122.

Charles, M. and Bogaard, A. (2005). Identifying livestock diet from charred plant remains: a Neolithic case study from Southern Turkmenistan. In J. Davies, M. Fabis, I. Mainland, M. Richards and R. Thomas (eds.) *Diet and health in past animal populations. Current research and future directions.* 9th ICAZ International Conference. Durham 2002, 93–103. Oxford, Oxbow Books.

Charles, M., Halstead, P. and Jones, G. (1998) The Archaeology of fodder: introduction. In M. Charles, P. Halstead and G. Jones (eds.) *Fodder: archaeobotanical, historical and ethnographic studies.* Oxford, Oxbow Books.

Copley, M. S., Berstan, R., Dudd, S. N., Docherty, G., Mukherjee, A. J., Straker, V., Payne, S. and Evershed, R. P. (2003) Direct chemical evidence for widespread dairying in prehistoric Britain. *Proceedings of the National Academy of Sciences* 100, 1524–1529.

Courty, M. A., Macphail, R. and Wattez, J. (1991) Soil micromorphological indicators of pastoralism; with special reference to Arene Candide, Finale Ligure, Italy. *Rivista de Studi Liguri* 57, 127–150.

Davies, J., Fabis, M., Mainland, I., Richards, M. and Thomas, R. (eds.) (2005) *Diet and health in past animal populations. Current research and future directions.* 9th ICAZ International Conference. Durham 2002. Oxford, Oxbow Books.

de Cupere, B., Lentacker, A., van Neer, W., Waelkens, M. and Verslype, L. (2000) Osteological evidence for the draught exploitation of cattle: first applications of a new methodology. *International Journal of Osteoarchaeology* 10, 254–267.

di Lernia, S. (2001) Dismantling dung: delayed use of food resources among early Holocene foragers of the Lybian Sahara. *Journal of Anthropological Archaeology* 20, 408–441.

Ertug-Yaras, F. (1996) Contemporary plant gathering in Central Anatolia: an ethnoarchaeological and ethnobotanical study. In M. A. Öztürk, Ö. Seçmen and G. Gök (eds.) *Plant life in Southwest and Central Asia.* Proceedings of the 4th Plant Life in Southwest Asia Symposium, 945–962. Izmir, Ege University Press.

Fontes, A. (2005) *Usos e costumes de Barroso.* Lisbon, Âncora Editora.

González Urquijo, J. E., Ibáñez Estévez, J. J., Zapata Peña, L. and Peña-Chocarro, L. (2001) Estudio etnoarqueológico sobre la cerámica Gzaua (Marruecos). Técnica y contexto social de un artesanato arcaico. *Trabajos de Prehistoria* 58, 5–27.

Hall, A. and Kenward, H. (1998) Disentangling dung: pathways to stable manure. *Environmental Archaeology* 1, 123–126.

Halstead, P. (1993) Banking on livestock: indirect storage in Greek agriculture. *Bulletin on Sumerian Agriculture* 7, 63–75.

Halstead, P. (1998) Mortality models and milking: problems of uniformitarianism, optimality and equifinality reconsidered. *Anthropozoologica* 27, 3–20.

Ibáñez Estévez, J. J., González Urquijo, J. E., Peña-Chocarro, L., Zapata, L. and Beugnier, V. L. (1999) Harvesting without sickles. Neolithic examples from humid mountain areas. In S. Beyries and P. Pétrequin (eds.) *Ethno-Archaeology and its transfers*, 23–36. Oxford, BAR International Series, 983.

Ibáñez, J. J., González Urquijo, J. E. and Moreno, M. (2001a) Le travail de la peau en milieu rural: le cas de la Jebala marocaine. In F. Audouin-Rouzeau and S. Beyries (eds.) *Le travail du cuir de la Préhistoire à nos jours.* XXIIᵉ rencontres internationales d'archéologie et d'histoire d'Antibes, 79–97. Antibes, Éditions APDCA.

Ibáñez, J. J., Peña-Chocarro, L., Zapata, L., González Urquijo, J. E. and Moreno-García, M. (2001b) Les récipients en bouse de vache et argile non cuite (*tabtoba* et *tonna*) dans la région Jebala (Maroc). Exemple d'un processus technique de type domestique. *Techniques et Culture* 38, 175–194.

Jones, G. (1998) Distinguishing food from fodder in the archaeobotanical record. *Environmental Archaeology* 1, 95–98.

Karkanas, P. (2006) Late Neolithic household activities in marginal areas: the micromorphological evidence from the Kouveleiki caves, Peloponnese, Greece. *Journal of Archaeological Science* 33, 1628–1641.

Kenward, H. and Hall, A. (1997) Enhancing bioarchaeological interpretation using indicator groups: stable manure as a paradigm. *Journal of Archaeological Science* 24, 663–673.

Korstanje, M. A. (2005) Microfossils in Camelid dung: taphonomic considerations for the archaeological study of agriculture and pastoralism. In T. O'Connor (ed.) *Biosphere to Lithosphere. New studies in vertebrate taphonomy.* 9th ICAZ International Conference. Durham 2002, 69–77. Oxford, Oxbow Books.

Lewthwaite, J. (1981) Plains tails from the hills: transhumance in Mediterranean archaeology. In A. Sheridan and G. Bailey (eds.) *Economic Archaeology*, 57–66. Oxford, BAR International Series, 96.

Lopes Dias, J. (1970). *Etnografia da Beira.* Lisbon, Câmara Municipal de Idanha-a-Nova.

Luik, H., Choyke, A. M., Batey, C. E. and Lõugas, L. (eds.) (2003) *From hooves to horns, from mollusc to mammoth. Manufacture and use of bone artefacts from Prehistoric times to the Present.* Proceedings of the 4th Meeting of the ICAZ Worked Bone Research Group at Tallinn, 26th–31th of August 2003. Tallinn, Muinasaja Teadus.

Millás Vallicrosa, J. M. (1943) El tratado de agricultura de Ibn Wafid. *Al-Andalus* 8, 281–332.

Millás Vallicrosa, J. M. (1948) La traducción castellana de "El tratado de agricultura" de Ibn Bassal. *Al-Andalus* 13, 347–430.

Millás Vallicrosa, J. M. and Aziman, M. (1955). *Libro de agricultura de Ibn Bassal.* Tetuán.

Miller, N. F. (1984) The use of dung as fuel: an ethnographic example and an archaeological application. *Paléorient* 10, 71–79.

Miller, N. F. and Smart, T. L. (1984) Intentional burning of dung as fuel: a mechanism for the incorporation of charred seeds into the archaeological record. *Journal of Ethnobiology* 4, 15–28.

Moreno-García, M. (1999) *The archaeozoology of transhumance in medieval Spain.* Unpublished PhD, University of Cambridge.

Moreno-García, M. (2004a) Hunting practices and consumption patterns in rural communities in the Rif mountains (Morocco) – some ethno-zoological notes. In S. J. O'Day, W. van Neer and A. Ervynck (eds.) *Behaviour behind bones. The zooarchaeology of ritual, religion, status and identity.* 9th ICAZ International Conference. Durham 2002, 327–334. Oxford, Oxbow Books.

Moreno-García, M. (2004b) Manejo y aprovechamiento de las cabañas ganaderas en las comunidades rifeñas marroquíes. *El Pajar. Cuaderno de Etnografía Canaria* 19, 84–90.

Mulville, J. and Outram, A. K. (eds.) (2005) *The Zooarchaeology of fats, oils, milk and dairying*. 9th ICAZ International Conference. Durham 2002. Oxford, Oxbow Books.

Pimenta, C. M. (2004) Las abejas: un recurso de gran valor y utilidad para las comunidades agro-pastoriles del Rif (Marruecos). *El Pajar. Cuaderno de Etnografía Canaria* 19, 91–95.

Rawat, D. S., Joshi, R. and Joshi, M. (2000) Indigenous methods for storage and use of bioresources: case study, Indian Central Himalaya. *AMBIO: A Journal of the Human Environment* 29, 356–358.

Scott, L. (1987) Pollen analysis of hyena coprolites and sediments from Equus Cave, Taung, southern Kalahari (South Africa). *Quaternary Research* 28, 144–156.

Shahack-Gross, R., Albert, R. M., Gilboa, A., Nagar-Hilman, O. *et al.* (2005) Geoarchaeology in an urban context: the uses of space in a Phoenician monumental building at Tel-Dor (Israel). *Journal of Archaeological Science* 32, 1417–1431.

Sillar, B. (2000) Dung by preference: the choice of fuel as an example of how Andean pottery production is embedded within wider technical, social, and economic practices. *Archaeometry* 42, 43–60.

Viegas Guerreiro, M. (1982). *Pitões das Júnias. Esboço de monografia etnográfica*. Lisbon, Serviço Nacional de Parques, Reservas e Património Paisagístico.

Viegas Guerreiro, M., de Abreu, D. and Ferreira, F. M. (1982). *Unhais da Serra. Notas geográficas, históricas e etnográficas*. Lisbon, Serviço Nacional de Parques, Reservas e Património Paisagístico.

Visscher, P. K., Vetter, R. S. and Robinson, G. E. (1995) Alarm pheromone perception in honey bees is decreased by smoke (Hymenoptera: Apidae). *Journal of Insect Behavior* 8, 11–18.

Wilson, D. G. (1984) The carbonisation of weed seeds and their representation in macrofossil assemblages. In W. van Zeist and W. A. Casparie (eds.) *Plants and Ancient Man*, 201–206. Rotterdam, Balkema,

Zapata Peña, L., Peña-Chocarro, L., Ibáñez Estévez, J. J. and González Urquijo, J. E. (2003) Ethnoarchaeology in the Moroccan Jebal (Western Rif): wood and dung as fuel. In K. Neumann, A. Butler and S. Kahlheber (eds.) *Food, fuel and fields. Progress in African Archaeobotany*, 163–175. Köln, Africa Praehistorica, 15.

5. Folk taxonomies and human–animal relations: The Early Neolithic in the Polish Lowlands

Arkadiusz Marciniak

This paper intends to draw attention to animal categorisation and classification as a significant aspect of human-animal relations, which aids in understanding different attitudes towards domestic and wild animals in the Neolithic. In particular, it will examine the heuristic potential of folk taxonomy in this regard and challenge the modern animal classification scheme in the tradition of Carl Linnaeus in zooarchaeology. It will also discuss the dominant theoretical standpoints of folk biology and taxonomy, including major issues of animal categorisation and classification. The main methodological issue of how to provide justified tools for getting to know the relations of the Neolithic farmers with their natural world, in particular with animals, will also be scrutinized. Lastly, the paper will aim at proposing a classificatory scheme of animals in the Early Neolithic in the Polish lowlands based upon a tradition offered by the ethnoscience school of folk taxonomy and will apply these concepts to interpret significant differences in the treatment of different species.

Keywords: Neolithic, folk taxonomy, classification, North European Plain, animals

Introduction

Zooarchaeological studies reveal different attitudes towards domestic and wild animals in the Early Neolithic. There are some striking parallels across various geographical zones. Early farmers treated different taxa in different ways, in particular sheep/goats and cattle. While the former was an ordinary source of meat, the latter was embedded in different social and ceremonial contexts. By no means can the early use of cattle be equated with meat-focused exploitation. The special significance of cattle was convincingly proved in various parts of the Near East, both in the Levant and Anatolia (*e.g.* Akkermans and Schwartz 2003, 75; Russell and Martin 2005), in the Balkans (Greenfield 2005, 28) and in the British Isles (Edmonds 1999, 28; J. Thomas 1999, 74). Similar differences are also discernible in the Early Neolithic in the Polish lowlands (Marciniak 2005).

There is a range of efficient and plausible zooarchaeological methods, which make it possible to discern how animals were treated, eaten, used in rituals, and utilized to secure and maintain social status, *etc*. While differences in the treatment of animals are often explicitly spelled out, causes of reported idiosyncrasies are rarely investigated and discussed. The aim of this chapter is to provide means for better understanding the very nature of observed differences in the treatment of particular animals during this early phase of farming groups' development.

People's actions are certainly conditioned, at least partly, by their way of knowing, understanding, and conceptualizing their world. The way people categorize various phenomena defines the manner in which they are perceived, interpreted and treated. Numerous ethnographic and ethnohistorical accounts make it clear that the system of animal classification employed has a profound impact upon the ways in which animals are maintained, tamed, and consumed. The position of animals in the classificatory scheme of the village communities of northeastern Thailand is responsible for dietary prohibitions including food taboo (Tambiah 1969). Classification of animals in New Guinea corresponds to their social categories, especially those related to ancestry and affinity (Douglas 1997, 45). The division of animals into clean and unclean in the Bible has a profound importance for treatment of various animals in this tradition. Similar examples can be multiplied. Hence, any attempt to delve into social and ideological meanings associated with animals by prehistoric groups must be preceded by an examination of their classificatory scheme.

Differences in the treatment of major domestic species in the Early Neolithic are striking, in particular because they refer to animals that are phylogenetically similar and close in scientific taxonomies. Hence, we may be sure that these differences are not embedded in modern scientific systematics. By questioning the heuristic potential of these modern categories in regard to observed idiosyncrasies, one needs to propose justified and manageable alternatives along with the new methods that may overcome the limitations of the old ones. In particular, these new methods need to be attentive to the nature of studied phenomena.

Modern classification of the natural world is only one of the many taxonomies invented by local communities, both in the past and today. These folk taxonomies vary to different degrees from the Linnaean system that marks the beginning of modern zoology. One would expect that prehistoric animal categorisations were closer to folk taxonomies than to the scientific ones. The classification and categorisation of animals is culture-specific and marks out the confines of the social and symbolic realm of human-animal relations. Criteria of animal classification are diverse and depend upon the animals' role in rituals, energy balance and economy cost-benefits, and in morphological traits such as size, shape and colour (Hunn 1982; Wapnish 1995). Hence, implementation of the folk taxonomy scheme into interpretation of human-animals relations in the past can challenge today's 'takens-for-granted' inherited from the contemporary classification schemes, and can bring forward more adequate categories intended to embrace the very role of animals in prehistory. Yet the importance of folk taxonomies is largely underestimated, and the Linnaean taxonomy is commonly and uncritically adopted in archaeological and zooarchaeological narratives.

In this chapter, I intend to draw attention to animal categorisation and classification as a significant aspect of human-animal relations that is often left unnoticed by contemporary zooarchaeology. Accordingly, this categorization should form the underpinnings of analysis of human-animal relations. I will examine the heuristic potential of folk taxonomy in this regard. The significance of a classification aiming at conceptualizing the perception of the world in general and the cognitive potential of recent developments in ethnobiology in archaeology in particular, has hardly been explored to date. It is clear, however, that no single categorization of the natural world made of relatively fixed categories can adequately describe different human-animal relations and the taxonomy schemes of different groups.

The main methodological issue is to provide justified tools for getting to know these intimate relations of the Neolithic farmers with their natural world, in particular with animals. Discerning the idiosyncratic folk taxonomies of different prehistoric groups through faunal remains is by no means an easy and straightforward task to complete. Any animal bone assemblage can be grouped into a number of categories based upon a range of criteria. For example, they may include domestic/wild, food/non-food, domestic production or trade (Serjeantson 2000). The problem arises in discerning which of those originate in the way in which animals were classified by prehistoric groups. During initial analysis, animal bones are certainly identified to scientifically defined species based upon known modern specimens. However, diverse patterns of animal treatment and use are being recognized at a higher level, assembling a certain range of species into separate categories such as clean vs. unclean, edible vs. inedible, or ceremonial vs. common. Different treatment of animals belonging to these high rank categories arguably results from the particular classificatory scheme in place.

In this chapter, I aim to discuss the heuristic potential of a folk taxonomy scheme in order to achieve better comprehension and understanding of human-animal relationships in the Early Neolithic in the Polish part of the North European Plain. The dominant theoretical standpoints of folk biology and taxonomy, as well as major issues of animal categorisation and classification raised by ethnoarchaeological and ethnohistorical studies, will be presented. This will be followed by an attempt to apply these concepts to interpret significant differences in the treatment of different species, in particular cattle and sheep/goat. The differences between cattle and pig, on one side, and sheep/goats on the other have been identified in terms of taphonomy, body part composition, and spatial distribution, as well as by association with other kinds of archaeological evidence which was statistically significant in all contexts throughout the studied settlements (see Marciniak 2005).

Classifications and Taxonomies of the Animal World

Classification Versus Taxonomy

The diversity of living organisms is overwhelming, which creates problems for humans who have a need to impose order on this complicated organic world that surrounds them. Hence, it is not surprising that humans spent lots of time and effort describing, naming and classifying this world and then building larger taxonomic schemes.

Classification can be understood as the practice aimed at defining and distinguishing groups or categories of organisms sharing certain similarities according to established set of criteria. Classificatory criteria can potentially be endless. Any given category needs to have a distinct set of features that is significantly distinct from other imposed categories. The process of classifying then means assigning an individual object to one of the distinguished categories. These categories are later arranged into a larger classificatory scheme in the course of a practice known as taxonomy.

Consequently, any taxonomic system is composed of distinctive categories, such as ranks (*e.g.* family, genus), that are mutually exclusive and unambiguous and arranged in a hierarchical manner. Taxonomies are then intrinsically typological. They form a kind of a tree structure for a given type of objects. Elements in such a system are ordered according to their presumed relationships, which may be different in scope. These categories are named, which often expresses the existing hierarchy between them. Taxonomies generally originate from local cultural and social contexts and serve various purposes.

The Linnaean Classification System

Modern classification systems for animals can be traced back to the Renaissance when scientists began to perceive the natural world as an entity, independent from humans, which has an existence of its own. As a result, animals began to be classified in a way that lacked a human-focused perspective – they were more detached and viewed independently (K. Thomas 1982).

This perspective was significantly strengthened with the groundbreaking classification of animals and plants by Carl Linnaeus (1707–1778), published in the 10th edition of his *Systema Naturae* in 1758. This publication marked the beginning of the modern zoological nomenclature. Linnaeus proposed a hierarchical system in which each organism belongs to a series of ranked taxonomic categories such as species, genus, family, order, class, phylum, and kingdom. The Linnaean taxonomy is a formal nested system of ranking organisms based on a simple hierarchical structure, from most general to most similar. At any given rank, each organism belongs to only one taxonomic group. Linnaeus also introduced the so-called binomial nomenclature system, whereby each species is referred to by a two-word name in Latin, consisting of the generic name followed by the specific name. It is worth noting, however, that the Linnaean taxonomy was not aimed at indicating evolutionary relationships between organisms, as the concept of evolution was unknown to Linnaeus and his contemporaries (see Medin and Atran 1999).

The Linnaean seven-layered hierarchical system is still used today as a tool for grouping organisms based on evolutionary history and for communicating the complexly interrelated products of evolution. Although our understanding of evolutionary relationships among organisms has greatly improved in the last century, it is by no means complete. Relationships among organisms, and groups of organisms, continue to be revised as new data becomes available.

The taxonomic-nomenclatural system, however, turned out to be, in many instances, insufficiently detailed. In the following centuries, not surprisingly, the system expanded and developed in different ways for different animal groups. As knowledge of the natural world progressed and the number of groups of organisms identified became larger and larger, it became necessary to create further subcategories such as tribe, division and cohort. Moreover, each category can also have prefixes to create a higher grouping (super-) or lower (sub-, infra-) subdivisions.

These developments led to the invention of numerous names for the same taxon as a result of scientists working independently in different languages and countries. Over time, it became clear that a universal system needed to be created in order to regulate the scientific naming of new animal taxa and to formulate methods for dealing with the myriad of synonyms that had already appeared. After a long delay, the formal international agreement on names and ranks laid out in the first edition of the *International Code of Zoological Nomenclature* was finally issued in 1961 (Stoll *et al.* 1961). The last edition, published in 1999, reflects the many contemporary voices of practising zoologists and serves as guidelines containing the rules of nomenclature (Ride *et al.* 1999).

However, the code of nomenclature is unable to accommodate the different schools, traditions, and personalities that affect how various taxa are organized. The major disagreement exists between two groups of taxonomists known as the *splitters* and the *lumpers*. Splitters make very small units by splitting genera among new families and putting species in new genera. Lumpers, on the other hand, prefer to lump together a large number of species in each genus, or genera in families, making large units. These different approaches for dividing up taxonomic units can create difficulties in establishing an appropriate name for some animals.

Modern taxonomy of animals in the Linnaean tradition is omnipresent in zooarchaeological studies, and the classification it advocates, along with its all consequences, is taken for granted and hardy debated. It is an outcome of modernist delimitation of cultural and natural domains into separate spheres. However, this modernist classification is one of many other taxonomies developed by local communities known as *folk taxonomies*.

Folk Taxonomies

Theoretical Traditions

In dynamically developing folk taxonomy, one can distinguish three main theoretical schools. The cognitive approach, in the tradition of Edmund Leach and Mary Douglas, concentrates on "how objects were named and how those names were grouped into larger units" (Wapnish 1995, 236). Symbolic anthropology, in the tradition of Claude Lévi-Strauss' structuralism, assumes that the classificatory system of all people formed a kind of over-reaching taxonomy, and that particular systems were closely related to each other (Lévi-Strauss 1963).

The third and most promising approach, however, is a perspective offered by ethnoscience. It is advocated by Brent Berlin and his associates, and appears in formalist and sociological variants. The ethnoscience approach is based upon the study of the principles of classification expressed and used by local groups in their own language. Hence, the recognition of any classification system is possible through study of the taxonomic relations along with the totality of lexical-semantic relations, including the phonology and grammar of local languages (Perchonock and Werner 1969, 237). Ethnoscience aims to spell out differences between the particular and universal ways in which different groups understand and conceptualise the plant and animal worlds.

In this perspective, the animal world is grouped into five categories:

(a) unique beginner;
(b) a life form understood as a group of organisms characterised by general morphological similarities, such as size, behaviour or habitat (*e.g.* fish, bird or mammal);

(c) a generic form referred to as "groupings of organisms in the natural environment" (Berlin *et al.* 1973, 215; see also Atran 1990), which is arguably the most common category in almost all folk taxonomies and comprises taxa defined by morphological, behavioural or psychological similarities;

(d) an intermediate form defined on the basis of habitat and/or morphology. These are, however, criteria which are far less specific than at the generic level. They are often rare and usually unlabelled;

(e) varietal form, which designates categories of significant cultural importance (Wapnish 1995, 248–249) and is also very rare in folk taxonomies.

Systematic studies of the folk classification of plants and animals by the formalist ethnoscience school revealed the existence of certain general 'regular structural principles' in folk taxonomies, despite obvious differences in their actual conceptualisations (Berlin *et al.* 1973, 214). In particular, a classification system is believed to be governed by four principles:

(a) enumeration of groupings of organisms of varying degrees of inclusiveness, similar to taxa;

(b) grouping of taxa into a small number of classes known as taxonomic-ethnobiological categories and defined by the linguistic and taxonomic criteria;

(c) hierarchical arrangement of these entities and taxa assigned to each rank are mutually exclusive, with exception of the unique beginner of which there is only one member;

(d) presence of the taxa of the same ethnobiological categories at the same level within any particular taxonomic structure.

Contrary to modern zoological systems, folk taxonomies are usually polynomic in nature, which results in delimiting paraphyletic taxa. This is a group of organisms defined upon shared primitive characteristics, as opposed to a monophyletic taxon, often defined by one or more uniquely shared characteristics. A range of features that justifies uniting various organisms as one category is usually very diverse and idiosyncratic. Consequently, the characteristics shared by the species of a paraphyletic taxon are often conspicuous. This is a result of three major factors:

(a) the relationships between 'practices' and the diversity of knowledge;

(b) the non-linearity of the relationship between 'prototype' (the paradigmatic classificatory focal center) and periphery, according to the logics of complex systems;

(c) the diversity of salience of the decontextualized ('abstract') category in different cultural worlds.

Various modes of ethnoclassification and nomenclature often require typification. However, this differs considerably from the collective efforts of modern zoology. It is always subjective as part of individual experience and practice. There are two opposite processes in classification in folk taxonomy: (a) a descending process of division of large, more inclusive classes into smaller, less inclusive ones; and (b) an ascending process of progressive abstraction and generalisation, by which lower level taxa become progressively grouped into higher level taxa. Folk generic names are basically the starting point of both processes, of subdivision into species and of abstraction into the higher categories.

These general schemes are proved to exist in animal folk classifications in many ethnohistorical and anthropological contexts, such as among the Karam and Fore (*e.g.* Bulmer and Tyler 1968; Diamond 1966) and Ndumba from the East New Guinea Highlands (Hays 1983). The Fore group developed an exhaustive system of classification consisting of two levels. The first comprises a higher category called 'big names'. It is further subdivided into lower units called 'small names.' Among 'big names', the Fore list both smaller flightless animals, such as rodents and marsupials, and large flightless creatures, such as echidna and giant rats (Diamond 1966).

However, other studies reveal some inconsistencies in Berlin's formalist approach. The most serious problem refers to the difficulty of assigning folk taxa to ranks, considering the complicated structure of the customary labels of particular taxa. A classificatory category, such as taxon, might have been recognised as a basic one for some people, while it could have been treated as a subordinate or superordinate for others. This depends upon the particular trajectory of a group's history and more particularly upon its knowledge, memory and culture (Mandler *et al.* 1991, 266).

Taxonomic Categories
The major taxonomic category of the animal world is arguably the generic species. These are commonly encountered among contemporary groups studied by folk taxonomies. They are divided in a "virtually exhaustive manner" (Atran 1999, 124) and are usually included into some life form. The rank of a generic species is distinguished by biological properties such as inheritance, growth, and psychological function, as well as other more 'hidden' properties (Atran 1999, 127), and is defined by their "manifestly visible dimensions of the everyday world, that is, to phenomenal reality" (Atran 1990, 3). They can be called 'phenomenal givens', which means that they are recognized in an empirical way (Hull 1999, 491). Some generic species are labelled as binomials that make hierarchical relation between the generic species and the life form (Atran 1999, 125).

Generic species often correspond to the scientific genus (Atran 1999, 125), but this relationship is less significant here and can hardly be grasped archaeologically. This relationship is reportedly not isomorphic, and it may vary between a number of scientific species in comparison to generic species. It is also reported to be sometimes significant in regions studied by folk taxonomists (Atran 1999, 126). Subordinate ranks of folk species and varietals correspond to variations recognized and appropriated by human groups and are culturally specific (Atran 1999, 128).

Intermediate taxa exist between the generic species and life-form levels. They are not as well delimited as these two forms and are usually unnamed but exceptions are also reported. Intermediate taxa do not constitute mutually exclusive and exhaustive categories and are usually formed at a level approximately placed between the Linnaean family and order (Atran 1999, 129).

The highest rank comprises life forms. Members of life forms are usually biologically diverse but often share a number of perceptual diagnostics including stem habit, skin covering, *etc.* Most life forms are named (Atran 1999, 122).

Reconstructing Taxonomies of Prehistoric Groups
One of the major problems for the application of the ethnoscience scheme to the past is that of methodology. We are certainly not in possession of a range of methods making possible an application of all facets of ethnobiological taxonomies. We certainly cannot know how these peculiar categories existed in the language of Neolithic groups. Also, the difference between ancient folk categories and contemporary scientific concepts, the nature of folk categories, and the extent to which folk categories were essentialist and normative are all beyond the grasp of archaeologists. So, while the theoretical framework is useful and justifiable for studying the past, the methods are not.

However, considering that the results of contemporary folk taxonomy studies equivocally imply that the animal classification scheme significantly determines the way in which domesticated animals are apprehended and used, any attempt to conceptualise the character of animal classification among Central European early farmers can certainly contribute to better understanding the ways in which animals were treated in the Neolithic. It means that zooarchaeologists need to think about different methods for applying this theoretical platform. This is particularly important because the prehistoric system of animal classification was closer to folk taxonomy than to the scientific one. While its detailed reconstruction is hardly possible, an attempt of application of folk taxonomy onto the past is certainly a worthwhile pursuit. Folk taxonomy as such has to "mediate between the zoological categories, the ancient terms, and the bones" (Wapnish 1995, 233).

We must strive to bring into the foreground more adequate categories of analysis – those that are intended to grasp the very role of animals in past societies. If the goal is to comprehend the folk taxonomies of the past, applicability of any formal taxonomic systems is limited. Hence, any search for universal cross-cultural rules is heuristically inefficient from the point of view of increasing our understanding of the behaviour of the past people as a consequence of the reported idiosyncrasies in human-animal relations. At the same time, application of any classificatory schemes recognised by folk biology to the prehistoric past is by no means easy and straightforward since "all classification[s] are discursive practices situated in a given social matrix and [the] general configuration of knowledge and ideas" (Ellen 1979, 17).

Hence, the most comprehensive folk taxonomy system seems to comprise a combination of certain formal 'principles' (*sensu* Berlin *et al.* 1973, 214), the "social context of classificatory activity" and inherited tradition defining "content of different semantic fields" (Ellen 1979, 18). As mentioned earlier, despite a number of idiosyncrasies, folk taxonomies reveal "regular structural principles" (Berlin *et al.* 1973, 214) and are characterised by a certain degree of logical similarities, which justifies their use for interpreting human-animals relations in the Central European Neolithic.

Considering that the animal classification scheme considerably determines the way in which domesticated animals are apprehended and used, we may expect to discern similarities and differences in the treatment of different species through a range of zooarchaeological methods, which may originate from the classificatory scheme of the animal world in the studied society. We know from folk biology that folk generic species more or less correspond to scientific species (Medin and Atran 1999, 8), and hence we should not expect to find idiosyncrasies at this rank. This is additionally justified as we identify generic species by analysing and classifying faunal remains into species level. As easy as the identification of a generic category might be, capturing idiosyncratic categorizations within this category proves to be very difficult. I would doubt that this kind of fine-tuning of a generic species category in archaeology is at all possible due to a range of analytical problems.

As mentioned earlier, folk-biological taxonomy is characterized by a stable hierarchy of inclusive groups of organisms (Atran 1999, 121). However, at the higher ranks, and in particular at the intermediate level rank, we know from contemporary folk taxonomies that correlations with taxa are not so tight and do not match the Linnaean scheme. Hence, I would argue that this is the very level of the classificatory scheme at which we should look while trying to shed some light into human-animals relations in the Early Neolithic.

Human–Animal Relations in the Past: An example from the Early Neolithic of the Polish Lowlands

Introducing the Early Neolithic of the Polish Lowlands
The Early Neolithic in the Polish part of the North European Plain can be used as a test for investigating the above approach. It is an area rich in data, with tight chronological controls and high quality recovery techniques. The area is also at the beginning of the entry of early farming groups into the North European Plain. It lays the foundation for the development of food producing societies across much of northern Europe.

Early farmers emerged in the North European Plain in the second half of the 6th millennium BC and continued uninterrupted development through the first half of the 5th millennium BC. They are represented by the Linear Band Pottery Culture communities. This region was colonized by immigrants from south-eastern Europe who brought with them a whole array of new material culture including longhouses, a simple style pottery with curvilinear and rectilinear motifs, and stone technology in the form of symmetrical axes and heavy adzes with a plano-convex cross section. They practiced mixed-farming subsistence technology. The Linearbandkeramik (LBK) culture covered large areas of Central Europe. In its earlier phases, it is characterized by remarkable uniformity over vast geographical distances, and its material culture was of limited stylistic variability in various regions. We have to bear in mind, however, that these communities were different in many respects from what is commonly taken for granted; namely, that communities consisted of settled people with domesticated herds of animals and small fields around their settlements – more or less how contemporary peasants are viewed (Marciniak 2008). It has been argued that the inhabitation of the new territories was accompanied by the construction of a communal identity and ancestry that involved mobilisation of external cultural resources (*sensu* Giddens 1987), including new ideas about housing and exotic resources, such as flint and cattle (see Marciniak 2000).

Zooarchaeological Evidence

Detailed studies of animal bone remains and their archaeological context from the Early Neolithic settlements of the Polish part of the North European Plain revealed striking differences between cattle, sheep/goats and pigs in taphonomic pattern, body part representation, spatial distribution and association with other kinds of archaeological evidence (Marciniak 2005). These statistically significant differences in all contexts throughout the studied settlements are indicative of considerably varied treatment of these animals at these settlements. The small number of pig bones makes it difficult to discern rules of pig treatment in more detail. However, a revealed pattern may imply some similarities with cattle, but one need to treat this conclusion with caution.

Cattle bones are the most abundant faunal remains in the Early Neolithic of Central Europe. Taphonomic analysis implies a very peculiar method of cattle marrow consumption. The bones were first roasted, broken and then the cooked marrow consumed. This kind of breakage is very common in the studied material, and can be recognized by the very characteristic way the cooked bones were broken, which produced jagged fractures usually oriented transversely to the bone's long axis, accompanied by hammer stone and anvil dents. The break surface was often burned with ash and accompanied by numerous scratches that are an indication of post fire breakage. Long bones were broken transversely at mid-shaft or near epiphyses. Mandibles were broken in a similar way, indicating that they were exposed to fire, probably roasted, and broken only after that. This kind of marrow consumption appears as a common and quite peculiar culinary practice of the early lowland farmers and might have had a discursive character. Interestingly, sheep/goat marrow, albeit not roasted, was also consumed on a daily basis.

Cattle and pig body part representation is characterized by a deliberate selection of certain anatomical segments, in particular skulls, scapulae, and axial segments, and marked by avoidance of their limbs. Early farmers perceived and used sheep and goats differently than cattle. Anatomical part distribution is characterized by a variable composition of highly processed anatomical parts, which implies that all of them were eaten.

Consumption of cattle meat and marrow was clearly regarded as appropriate in one social context and inappropriate in another. This is indicated by the deposition of cattle bones in specific locales at the settlement, in particular in the open space between longhouses. The food was probably cooked in a hearth or oven located also outside the longhouses. The remains of communal consumption were deposited exclusively in the so-called clay pits located between longhouses and do not appear in other types of pits used at these settlements. This fixed practice also implies that the settlement space was categorized, perceived, and used by its inhabitants in a very straightforward way. Contrary to cattle consumption, that of sheep/goat took place in the house and/or directly around the house. The small number of pig bones recovered has made their spatial distribution analysis hardly conclusive.

The available evidence from the Early Neolithic settlements from the Polish part of the lowlands implies practicing at least two kinds of consumption among local farmers. The first focused on cattle while the other on sheep/goats. Fragmentary evidence implies that pigs were also an important element in feasting and pork was not consumed on a daily basis.

Considering the small degree of cattle bone fragmentation in addition to the peculiar body part representation, this may imply communal ceremonial consumption which was performed in a very standardized and repeatable manner over the long period of time. This consumption was performed on a regular basis in the same way throughout the whole region, which is manifested by an identical deposition of cattle bones at different settlements over a considerable time span, as indicated by the same anatomical composition of cattle bones in all layers of deep pits.

The second kind of consumption had an ordinary character and comprised mainly of sheep/goats. The body part representation differs largely from that of cattle, and is characterized by a variable composition of all anatomical parts indicating that all animal parts were eaten. This kind of sheep/goats meat consumption usually took place in the house and/or directly around the house, and bone remains were deposited in pits around the entrances of these houses. No roasted marrow of sheep/goats was consumed.

Folk Taxonomies and Human–Animal Relations in the Early Neolithic of the Polish Lowlands

The classification scheme of the animal world used by the early farming groups of the Polish lowlands needs be set in the context of the regional trajectory of development. The occupation of the lowlands had been accompanied by the construction of the early farmers' communal identity and ancestry, and maintenance of their security in a new 'frontier' situation and unknown environment. This may have involved the mobilisation of external cultural resources such as the idea of housing, exotic flint and copper. Domesticated animals, especially cattle, were arguably also one of these important social and symbolic resources which provided metaphors for the creation of the group and its communal identity. Animals, therefore, along with a set of associations they were intertwined with, are to be viewed as "vehicles of social relations" (Ingold 1984, 7). The importance of cattle was then built in the continuing process of moving farmers northwards into new areas, which was accompanied by social fragmentation that became an intrinsic element of these communities. Cattle accompanied farmers in this dispersal and under these circumstances local communities used "the potential of cattle to build enduring social bonds" (Comaroff and Comaroff 1991, 37). For groups that successfully colonized these new areas and were then living in a rather hostile environment and relative isolation, cattle were the very basis for maintaining their stability and provided security. Cattle became "cultural objects" incorporating an "extended" form of signification (Giddens 1987, 100). They may have been an embodiment of the past, history, and memory, and may have become a focus of meaning in which social co-operation and practice were undertaken. This may have been accompanied by an expansion of invented terms, which is reported as a characteristic phenomenon during the emergence of farming (Brown 1985). Early farmers may have used a binomial nomenclature. The existence of names is indicative of knowledge, not only of organism morphology and behaviour, but also of use, including the usefulness of particular parts (Ellen 1999, 104–106).

As a result, Early Neolithic farmers had peculiar knowledge profiles and their important element was ethnobiological knowledge. As argued by Ellen (1979), any classification is always created within certain social conditions and circumstances, which can be divided into situational and structural ones. The former refers to the specific, not-yet-fixed social context in which classificatory activity is the process of creating, while the latter refers to stabilised social and economic relations. I would argue that conceptualization of the animal world by early farmers was based upon substantive knowledge, which is the knowledge that people apply when engaged in relation to animals and plans (see Ellen 1999), which may have been encoded and registered in their language. Animal categorization in the Early Neolithic in the North European Plain is hence the result of situational knowledge which was far more significant than perceptual criteria, and "this kind of knowledge is the result of generations of accumulated experience, experimentation and information exchange." This kind of deductive models of "how the natural world works" is then more significant than accumulated inductive knowledge (Ellen 1999, 106).

Like many other farming groups, Early Neolithic people probably had a tendency to classify animals hierarchically, so as to make order in the world. Generic species level was arguably the most informative one. They probably followed situational social conditions, defined by their existence in newly colonized territories, which structured their way of classification. They may be well described in Ellen's (1979, 18) words as an "intracultural variant determined by either a specific social context of classificatory activity or material or perceived content of different semantic fields" (Ellen 1979, 18). The basic building blocks of this early Neolithic taxonomy may have been generic taxa such as cattle, sheep/goat or pig. There is no reason to predict that these were perceived and categorised as different categories bearing their morphological and behavioural distinctiveness.

These lower-level generic categories are believed to have existed earlier, both conceptually and epistemologically. Once these are distinguished, they had to be defined in order to be grouped at a higher level. The grasped similarities and differences between generic species resulted in their being placed in different higher categories. Thus, I would argue that these taxa may have been grouped differently than expected in scientific taxonomy, in particular, at the intermediate rank level.

These new intermediate forms were created as a manifestation of a general tendency among farming groups to create ranks of folk-specific and varietals that are significantly more common than among non-farmers (Ellen 1999, 109). Intermediate taxa among early farmers were arguably created between subfamily levels of Bovinae and Caprinae, family Suidae and order Artiodactyla. As implied by the distinctly specific ways of treatment of particular species identified by taphonomic patterns, body part representation, spatial distribution as well as depositional practices revealed by their association with other kinds of archaeological evidence, cattle and pig might have belonged to one intermediate taxon while sheep/goat to the other. Consequently, the first might have been named *animals-recently-domesticated*, while the latter *animals-already-domesticated*. Whatever the real names were, we can be almost sure that it clearly articulated this classification, bearing in mind what Berlin *et al.* (1973, 216) observed on the basis of contemporary studies, namely that "nomenclature is often a near perfect guide to folk taxonomic structure."

The assignment of generic taxa to higher ranks was certainly conducted on the basis of biological as well as cultural and social factors (see Berlin *et al.* 1973). The latter comprised the remembered and current importance of particular taxa and their shared significance, going beyond a simple distinction between particular species.

This significant distinction may have referred to the social memory of the farming groups colonising the North European Plain from the south. Neither sheep nor goats had their ancestors living among the wild animals of temperate Europe (Glass 1991, 30), while cattle and pigs may have been locally domesticated in the Polish part of the lowlands, albeit the degree of local domestications may vary in different parts of Central Europe (*e.g.* Götherström *et al.* 2005; Bollongino *et al.* 2006). Hence, locally domesticated animals such as cattle and pigs in the Polish lowlands may have been perceived and classified differently than those which were originally part of the Balkan Early Neolithic package. While the DNA analyses of southern European samples are pointing toward ultimate ancestry of domestic cattle and pigs in the Near East, none of the Polish lowlands material has of yet been successfully tested. This may imply local trajectory of development of the Neolithic groups migrating to the area from the southeast of Europe. Clearly, this classification scheme was further responsible for the ways in which particular taxa were maintained, consumed, and the refuse disposed. Animal categorisation was obviously cultural and social in nature and was remembered, repeated and transformed in oral tradition.

Final Remarks

Zooarchaeological studies reveal striking differences in perception, treatment and use of a range of domestic and wild animals. They often refer to animals that are phylogenetically similar. Hence, we may predict that these differences are not embedded in modern scientific systematics. At the same time, numerous anthropological accounts make it clear that the employed animal classification scheme considerably determines the way in which animals are apprehended and used, which means that we may justifiably argue that discerned similarities and differences in treatment of different species may originate from the classificatory scheme of the animal world.

Since the modern Linnean taxonomic scheme may be too heuristically inefficient to grasp these idiosyncracies in human-animals relations, application of the folk taxonomy framework that proved to be efficient in recognizing the principles of classification in contemporary communities is required. Despite the fact that not a full range of methods of folk taxonomy can certainly be applied on archaeological and archaeolozoological materials from the past, an attempt to make use of this theoretical perspective is certainly a worthwhile pursuit.

The preliminary results of application of a folk taxonomy scheme for understanding apparent idiosyncrasies in human-animal relations in the Early Neolithic in the Polish lowlands may indicate that grouping major domesticated taxa on the intermediate form level may have shaped treatment of animals by these communities and defined their role in the process of inhabiting vast areas of the then unoccupied areas of the North European Plain.

Interestingly, the distinctive difference in treatment of major domesticated species is also well reported from Early Neolithic settlements in the Near East and Europe. Hence, one could predict that the particular significance of some animals, in particular cattle, was sanctioned by their role in the Near East and/or the Balkan Neolithic, which led to their dominant position in the classification scheme (see *e.g.* Friedberg 1979, 96–97) and nomenclature. Cattle may have achieved this position long before it was biologically domesticated. They were clearly of considerable ceremonial and symbolic importance and contributed only negligibly to the diet of the local inhabitants. This was primarily caused by social reasons, such as a means to create and maintain a group's identity, and to some extent also for ecological reasons. These, rather than supposedly economic effectiveness understood in terms of primary vs. secondary products, were the most important factors that influenced the ways in which animals were classified and exploited in the Early Neolithic in Central Europe. This broader picture, however, needs to be explored in far more details.

Accordingly, this significant distinction between major domesticated animals in the Polish lowlands may have referred to social memory of the farming groups colonising the North European Plain from the south. It was accompanied by the remembered and current importance of particular taxa and their shared significance beyond a simple distinction between particular species. This may have also involved local domestications of cattle and/or pigs. It is then justified to claim that the established classification of the animal world among early farming communities proved to be responsible for pretty standardized treatments of particular animals.

As a result of this inherited classificatory scheme, major domesticated animals were treated differently by these local farmers in the North. The importance of cattle for these groups was clearly much larger than providing meat or milk. They probably signified ties and stressed relations with living and previous generations. This special significance of cattle was manifested in communal feasting that may have been a part of tradition and memory, which were brought into being in everyday practice to a large extent through material objects such as houses or other monuments. Pigs also held a significant and complex position that went far beyond simple provision of meat for everyday consumption. They were possibly a substantial component of social practices linked to the communal identity. At the same time, early farmers perceived and used sheep and goats, exploited mainly as a source of meat for everyday consumption, differently than cattle and pigs.

In any case, considering the significant impact of classificatory schemes upon human–animal relations, we cannot leave aside the results of ethnobiology studies despite a range of methodological difficulties. They open up a new and fascinating avenue to be explored for the years to come.

Acknowledgements

I would like to take this opportunity to thank Umberto Albarella for his useful remarks and for compassionately guiding me through the editorial process. I am particularly grateful to Haskel J. Greenfield for his thorough comments on the earlier draft of this chapter. Comments by two anonymous reviewers helped to clarify certain issues discussed here.

References

Akkermans, P. M. M. G. and Schwartz, G. M. (2003) *The Archaeology of Syria. From Complex Hunter-Gatherers to Early Urban Societies (ca. 16,000–300 BC)*. Cambridge, Cambridge University Press.

Atran, S. (1990) *Cognitive Foundations of Natural History. Towards an Anthropology of Science*. Cambridge, Cambridge University Press.

Atran, S. (1999) Itzaj Maya folkbiological taxonomy. Cognitive universals and cultural particulars. In D. L. Medin and S. Atran (eds.) *Folkbiology*, 119–203. Cambridge, The Massachusetts Institute of Technology.

Berlin, B., Breedlove, D. E. and Raven, P. H. (1973) General principles of classification and nomenclature in folk biology. *American Anthropologist* 75, 214–242.

Bollongino, R., Edwards, C. J., Alt, K. W., Burger, J. and Bradley, D. G. (2006) Early history of European domestic cattle as revealed by ancient DNA. *Biology Letters* 2, 155–159.

Brown, C. H. (1985) Mode of subsistence and folk biological taxonomy. *Current Anthropology* 26, 43–53.

Bulmer, R. N. and Tyler, M. (1968) Karam classification of frogs. *Journal of Polynesian Society* 77, 333–385.

Comaroff J. and Comaroff, J. L. (1991) "How beasts lost their legs", Cattle in Tswana economy and society. In J. G. Galaty and P. Bonte (eds.) *Herders, Warriors, and Traders*, 33–61. Boulder, Westview Press.

Diamond, J. M. (1966) Zoological classification system of a primitive people. *Science* 151, 1102–1104.

Douglas, M. (1997) Deciphering a meal. In C. Counihan and P. van Esterik (eds.) *Food and Culture. A Reader*, 36–54. London and New York, Routledge.

Edmonds, M. (1999) *Ancestral Geographies of the Neolithic. Landscapes, Monuments and Memory*. London and New York, Routledge.

Ellen, R. (1979) Introductory essay. In R. Reason and D. Ellen (eds.) *Classifications in Their Social Context*, 1–32. New York, Academic Press.

Ellen, R. (1999) Models of subsistence and ethnobiological knowledge: Between extraction and cultivation in southeast Asia. In D. L. Medin and S. Atran (eds.) *Folkbiology*, 91–118. Cambridge, The Massachusetts Institute of Technology.

Friedberg, C. (1979) Socially significant plant species and their taxonomic position among the Bunaq of Central Timor. In R. Reason and D. Ellen (eds.) *Classifications in their Social Context*, 81–99. New York, Academic Press.

Giddens, A. (1987) *Social Theory and Modern Sociology*. Cambridge, Polity Press.

Glass, M. (1991) *Animal Production Systems in Neolithic Central Europe*. BAR International Series 572. Oxford, Archaeopress.

Götherström, A., Anderung, C., Hellborg, L., Elburg, R., Smith, C., Bradley, D. G. and Ellegren, H. (2005) Cattle domestication in the Near East was followed by hybridization with aurochs bulls in Europe. *Proceedings: Biological Sciences* 272, 2345–2350.

Greenfield, H. J. (2005) A reconsideration of the secondary products revolution, 20 years of research in the central Balkans. In J. Mulville and A. Outram (eds.), *The Zooarchaeology of Milk and Fats*, 14–31. Oxford, Oxbow.

Hays, T. E. (1983) Ndumba folk biology and general principles of ethnobiological classification and nomenclature. *American Anthropologist* 85(3), 592–611.

Hull, D. L. (1999) Interdisciplinary dissonance. In D. L. Medin and S. Atran (eds.) *Folkbiology*, 477–500. Cambridge, The Massachusetts Institute of Technology.

Hunn, E. (1982) The utilitarian factor in folk biological classification. *American Anthropologist* 84(4), 830–847.

Ingold, T. (1984) Time, social relationships and the exploitation of animals: Anthropological reflections on prehistory. In J. Clutton-Brock and C. Grigson (eds.) *Animals and Archaeology 3. Early Herders and their Flocks*. BAR International Series 202, 3–12. Oxford, Archaeopress.

Lévi-Strauss, C. (1963) *Structural Anthropology*. New York, Basic Books.

Linnaeus, C. (1758) *Systema Naturae* (10th edition). Holmiae [Stockholm], Laurentii Salvii.

Mandler, J. M., Bauer, P. J. and McDonough, L. (1991) Separating the sheep from the goats. Differentiating global categories. *Cognitive Psychology* 23, 263–298.

Marciniak, A. (2000) Living space. Construction of social complexity in Central European communities. In A. Richie (ed.) *Neolithic Orkney and its European context*, 333–346. Cambridge, McDonald Monographs in Archaeology.

Marciniak, A. (2005) *Placing animals in the Neolithic. Social Zooarchaeology of Prehistoric Farming Communities*. London, UCL Press.

Marciniak, A. (2008) Interactions between hunter-gatherers and farmers in the Early and Middle Neolithic in the Polish part of the North European Plain. In D. Papagianni, H. Maschner and R. Layton (eds.) *Time and Change. Archaeological and Anthropological Perspectives on the Long-Term in Hunter-Gatherer Societies*, 115–133. Oxford, Oxbow.

Medin, D. L. and Atran, S. (1999) Introduction. In D. L. Medin and S. Atran (eds.) *Folkbiology*, 1–15. Cambridge, The Massachusetts Institute of Technology.

Perchonock, N. and Werner, O. (1969) Navaho systems of classification: Some implications for ethnoscience. *Ethnology* 8(3), 229–242.

Ride, W. D. L., Cogger, H. G., Dupuis, C., Kraus, O., Minelli, A., Thompson, F. C. and Tubbs, P. K. (eds.) (1999) *Internation Code of Zoological Nomenclature* (4th edition). London, International Trust for Zoological Nomenclature.

Russell, N. and Martin, L. (2005) Çatalhöyük mammal remains. In I. Hodder (ed.) *Inhabiting Çatalhöyük: Reports from the 1995–99 seasons*. Cambridge, McDonald Institute for Archaeological Research.

Serjeantson, D. (2000) Good to eat and good to think with. Classifying animals from complex sites. In P. Rowley-Conwy (ed.) *Animal Bones, Human Societies*, 179–189. Oxford, Oxbow.

Stoll, N. R., Dollfus, R. Ph., Forest, J., Riley, N. D., Sabrosky, C. W., Wright, C. W. and Melville, R. V. (eds.) (1961) *International Code of Zoological Nomenclature adopted by the XV International Congress of Zoology (1st edition)*. London, International Trust for Zoological Nomenclature.

Tambiah, S. J. (1969) Animals are good to think and good to prohibit. *Ethnology* 8(4), 424–459.

Thomas, J. S. (1999) An economy of substances in earlier Neolithic Britain. In J. Robb (ed.) *Material Symbols. Culture and Economy in Prehistory*, 70–89. Carbondale, Southern Illinois University Press.

Thomas, K. (1982) *Man and the Natural World. A History of Modern Sensibility*. New York, Pantheon Books.

Wapnish, P. (1995) Towards establishing a conceptual basis for animal categories in archaeology. In D. B. Small (ed.) *Methods in the Mediterranean, Historical and Archaeological Views on Texts and Archaeology*, 233–273. Leiden, New York and Köln, E. J. Brill.

Fishing, Hunting and Foraging

6. The historical use of terrestrial vertebrates in the Selva Region (Chiapas, México)

Eduardo Corona-M and Patricia Enríquez Vázquez

This paper compares two diachronic sources of data, archaeozoological and ethnozoological, to develop hypotheses concerning the changes and persistence in the uses of faunal resources in the Selva region of Chiapas, México. Vertebrates had three main uses: food, raw material, and medicine. In order to discern the use of faunal resources in each chronological period, the data were processed by a taxonomic index and by two multivariate analyses. Using a global list of 64 terrestrial vertebrates, these data show that 13% of these animals are not currently used, but 45% of these vertebrates have been used from the Pre-Hispanic period to modern times. This comparison offers an initial hypothesis about the uses, shows the importance of local resources, and suggests the influence of cultural attitudes toward animals in a limited geographic scenario.

Keywords: México, Chiapas, zooarchaeology, ethnoarchaeology, ethnozoology

Introduction

Wild fauna is an essential component of the life of many human populations. In Mesoamerican cultures, diverse relationships between humans and animals were maintained. Animals were primarily used as food and raw materials, as well as in symbolic or ideological interactions (Corona-M and Arroyo-Cabrales 2003) (Figs. 6.1 and 6.2). These relationships were, and still are, linked to

Fig. 6.1 Some current mammal uses in the studied area: Bag with armadillo skin. Photo by Patricia Enríquez.

Fig. 6.2 Some current mammal uses in the studied area: Mousses dish in a religious offering. Photo by Patricia Enríquez.

faunal availability and cultural attitude; however, those factors were not static through time. In order to evaluate diachronic changes and persistence in the use of vertebrates, we propose the broad comparison of archaeozoological and ethnozoological data. It is remarkable to us that these types of specific historical and regional comparisons are scarce, though recently the issue has raised the interest of some researchers (Harris 2006).

The Mexican State of Chiapas currently possesses both the highest ethnic diversity as well as biodiversity of the country (CONABIO 1998; Toledo *et al.* 2002). Despite the fact that most of the archaeological research in Chiapas dates from the 19th century, it was not until recent decades that the importance of animal resources for human subsistence was recognized (Green and Lowe 1967; Chavez-O. 1969; Flannery 1969; Voorhies 1976; García-Bárcena 1982; Marrinan 1986; Martínez-Muriel 1989), particularly for the period between the Archaic and the Late Post-Classic eras (following the chronology used by Emery [2004a]). Additionally, ethnobiological research in Chiapas has mainly focused on botanical data, with studies concerning the use of animals by modern indigenous peoples being neglected (Enríquez *et al.* 2006). This paper uses a diachronic data set and focuses on terrestrial vertebrates in order to offer more reliable hypotheses concerning changes as well as persistence in the uses of animal resources in the Selva Region of the State of Chiapas (México).

Materials and Methods

The Selva region, as defined here, is a Mexican administrative division that comprises 14 municipalities in an area of almost 20,000km² (Fig. 6.3). The main geographic feature of this region is a mountainous area, with altitudes of nearly 2000m above sea level. The climate varies with the altitude, from humid and tropical in the lowlands to temperate in the highlands. In the past, the vegetation was tropical rain forest and mixed pine-oak forest; most of the area has been deforested due to increasing human disturbance as well as the predominance of *milpas*, coffee plantations, and grasslands (Gobierno del Estado de Chiapas 2005).

Several localities with archaeozoological material were chosen from the Selva and surrounding regions to represent the timeframe from the Pre-Ceramic through the Post-Classic periods. Unfortunately, we could not locate a reliable source concerning the use of fauna from the Colonial period in the study area or within the State of Chiapas. For the Pre-Ceramic period, no appropriate Archaic period locality was found within the study area; however, to cover this gap, the locality of Santa Marta Rock Shelter was selected. This site dates between 6400–4000 BC and demonstrates a hunter-gatherer subsistence strategy (MacNeish and Peterson 1962; Álvarez 1976; García-Bárcena 1982; García-Bárcena and Santamaría 1982). For the Classic and Post-Classic periods, three localities from the Selva region are used in this study: Palenque (Ocaña 1997), Yaxchilán (Soto 1998), and Toniná (Álvarez *et al.* 1980; Álvarez 1982; Ocaña and Polaco 1982; Álvarez *et al.* 1990; Ocaña and Polaco 1990). These sites were grouped together for the Archaeozoological period. The Ethnozoological period data were obtained by ethnographic fieldwork, namely, 68 interviews with local people from four towns: Palenque, Loma Bonita, Flor de Marquez, and Lacanja Chansayab (Enríquez 2005), as well as use of other published ethnographic sources (Góngora-Arones 1987; Zolla 1994; Aranda 1997; Naranjo *et al.* 1997; March 1997; Guerra 2001; Cupul-Magaña 2003; Naranjo *et al.* 2004).

We agree with Emery (2004b) on the importance of clarifying the methods used in faunal analysis in order to allow researchers to compare samples and combine data sets and provide stronger and broader interpretations. This is, however, an ideal situation, and although we expect that in the near future better data sets will be available, in the current analysis we could only use published resources in developing initial hypotheses on the human-fauna relationships. In this case, we consider that the faunal lists from the analyzed sites are the common denominator in all of these data sources. However, these lists must be used judiciously, as the data deal with different analytical problems and strategies. Likewise, these data are affected by many processes such as geographical distribution of the species, taphonomic factors, recovery techniques, and analyst bias in faunal identification, both in ethnozoological and archaeozoological analyses. Some of these effects are more influential within specific faunal groups, especially amphibians, reptiles and birds.

The reliability of the vertebrate data in each locality was measured and compared by the taxonomic distinctness index (Clarke and Warwick 1998). This test is based on the comparison of taxonomic lists (samples), which are, in this case, based on five taxonomic levels: class, order, family, genus and species. This index is independent of sample

Fig. 6.3 Map of the State of Chiapas showing the Selva region (dark grey). Localities mentioned in the paper are marked as of the Pre-Ceramic (PC) or Archaeozoological (AZ) periods.

size and can be based on binary data, such as those used in this study. The localities selected were processed by a statistical bootstrap method in order to obtain confidence intervals at 95%. All the localities inside of these confidence intervals could be subject to further comparative analysis, as indicated below. Also, in this case the index result suggests that the identified fauna in each locality was the product of use by local people.

Scientific names follow the checklists of Flores-Villela (1993) for herpetofauna, Howell and Webb (1995) for birds, and Reid (1997) for mammals. We categorized human use of fauna into three broad categories: food resources, medicine, and raw materials used in various manufacturing activities. These categories allowed comparison between faunal groups and chronologies.

The faunal lists were also processed as binary matrices in order to apply two multivariate analyses: unconstrained seriation and clustering. Seriation provides an ordering that can be interpreted as evidence of common and exclusive taxa for each chronology and locality, while cluster analysis finds similarities between localities. Using cluster analysis, distance is measured based on the Simpson index and defined as M/Nmin, where M is the total number of matches (a match is counted for all taxa present in both localities being compared). Nmin is the minimum number of taxa present in the two localities. We adopted this statistic tool because it detects shared taxa and treats two localities as identical if one is a subset of the other, thereby making it useful for fragmentary data (Hammer et al. 2007). The cluster groups are based on the algorithm UPGMA (Unweighted Pair-Group Moving Average). The results were measured from zero (no similarity) to one (total similarity). All the analyses were processed by PAST software (Hammer et al. 2007).

Results and Discussion

A global list of 64 species was obtained: 32 mammals, 15 birds, 12 reptiles, and five amphibians (Table 6.1)

Class	Order	Family	Identified taxon	Common name	Chronological stages	Uses
Amphibia		Bufonidae	*Bufo* sp	Toad	AZ	
	Anura	Hylidae	*Hyla* sp	Tree frog	EZ	M
			Smilisca baudini	Mexican tree frog	EZ	F
		Ranidae	*Rana* sp	Frog	AZ, EZ	M,F
		Rhynoprynidae	*Rhinophrynus dorsalis*	Mexican burrowing toad	EZ	F
Reptilia	Squamata	Corytophanidae	*Basiliscus vittatus*	Brown Basilisk	EZ	F
			Corytophanes sp	Helmeted	EZ	F
		Iguanidae	*Ctenosaura* sp	Black iguana	PC, EZ	F
		Polychridae	*Anolis* sp	Anole	AZ, EZ	M
		Teiidae	*Ameiva undulata*	Rainbow ameiva	EZ	M,F
	Serpentes	Boidae	*Boa constrictor*	Boa	EZ	F
		Chelydridae	*Chelydra serpentina*	Snapping turtle	EZ	F,R
		Viperidae	*Crotalus durissus*	Rattlesnake	EZ	M
	Testudines	Dermatemyidae	*Dermatemys mawii*	Mesoamerican river turtle	AZ, EZ	M,F,R
		Emydidae	*Trachemys scripta*	Common slider	PC, AZ, EZ	F,R
		Kinosternidae	*Kinosternon* sp	Musk turtles	PC, AZ, EZ	F,R
	Crocodylia	Crocodylidae	*Crocodylus moreletii*	Morelet's crocodile	AZ, EZ	M,F,R
Aves	Pelecaniformes	Sulidae	*Sula* sp	Brown booby	EZ	R
	Ciconiiformes	Ardeidae	*Ardea alba*	Great egret	EZ	R
	Ciconiiformes	Cathartidae	*Coragyps atratus*	Black vulture	EZ	M
	Falconiformes	Falconidae	*Falco* sp	Falcon	AZ	
		Accipitridae	*Buteo* sp	Roadside hawk	AZ, EZ	R
	Galliformes	Cracidae	*Ortalis* sp	Chachalaca	AZ, EZ	F
			Penelope purpurascens	Crested guan	PC, EZ	F
		Odontophoridae	not identified	Quails	AZ, EZ	F
			Crax rubra	Great curassow	PC, AZ, EZ	F
			Ara macao	Scarlet macaw	EZ	R
		Phasianidae	*Meleagris gallipavo*	wild turkey	AZ, EZ	F
	Psittaciformes	Psittacidae	*Amazona* sp	Parrots	EZ	M

Table 6.1 (continued overleaf) Taxonomical list of terrestrial vertebrates used in this analysis. Chronological stages: PC, Pre-Ceramic; AZ, Archaeozoological; EZ, Ethnozoological. Uses: M, medicine; F, food; R, raw material.

Class	Order	Family	Identified taxon	Common name	Chronological stages	Uses
Mammalia	Apodiformes	Trochiilidae	not identified	Hummingbirds	EZ	R
	Trogoniformes	Trogonidae	*Pharomachrus mocinno*	Resplendent quetzal	EZ	R
		Momotidae	*Momotus momota*	Blue-crowned motmot	EZ	R
	Didelphiomorpha	Didelphidae	*Didelphis* sp	Opossum	AZ, EZ	M
	Xenarthra	Dasypodidae	*Dasypus novemcinctus*	Nine-banded armadillo	PC, AZ, EZ	M, F
	Primates	Cebidae	*Allouatta* sp	Howler monkey	AZ, EZ	M, F
			Ateles geoffroyi	Spider monkey	AZ, EZ	F
	Rodentia	Sciuridae	*Sciurus* sp	Squirrel	PC, AZ, EZ	F
		Geomydae	*Orthogeomys* sp	Pocket gopher	AZ, EZ	M
		Heteromydae	*Liomys* sp	Spiny pocket mouse	PC	
			Heteromys sp.	Pocket mouse	AZ	
		Muridae	*Oryzomys* sp	Rice rat	PC, AZ	
			Sigmodon hispidus	Cotton rat	PC, AZ	
			Tylomys sp	Climbing rat	AZ	
			Ototylomys sp	Big-eared climbing rat	PC, AZ	
			Neotoma mexicana	Wood rat	AZ	
			Baiomys musculus	Southern pygmy mouse	PC	
			Reithrodontomys sp	Harvest mouse	PC, AZ	
			Peromyscus sp	White-footed mouse	PC, AZ, EZ	M
		Erethizontidae	*Coendou mexicanus*	Mexican porcupine	EZ	M, F
		Dasyproctidae	*Dasyprocta punctata*	Central american agouti	AZ, EZ	F, R
		Agoutidae	*Aguti paca*	Lowland paca	PC, AZ, EZ	F
	Lagomorpha	Leporidae	*Sylvilagus* sp	Cottontail rabbit	PC, AZ, EZ	F
	Carnivora	Procyonidae	*Procyon lotor*	Northern raccoon	EZ	F
			Nasua narica	Coati	PC, EZ	M, F
		Mustelidae	*Mephitis macroura*	Hooded skunk	EZ	M
			Spilogale putorius	Spotted skunk	EZ	M
			Conepatus mesoleucus	Hog-nosed skunk	AZ, EZ	M
		Felidae	*Puma concolor*	Cougar	AZ	
			Panthera onca	Jaguar	AZ, EZ	M, F, R
	Perissodactyla	Tapiridae	*Tapirus bairdii*	Tapir	AZ, EZ	F
	Artiodactyla	Tayassuidae	*Tayassu tajacu*	Collared peccary	AZ, EZ	F
			Dicotyles pecari	White-lipped peccary	PC, AZ, EZ	F
		Cervidae	*Odocoileus virginianus*	White-tailed deer	PC, AZ, EZ	F, R
			Mazama americana	Red brocket	PC, AZ, EZ	F, R

Table 6.1 (continued from previous page) Taxonomical list of terrestrial vertebrates used in this analysis. Chronological stages: PC, Pre-Ceramic; AZ, Archaeozoological; EZ, Ethnozoological. Uses: M, medicine; F, food; R, raw material.

were present in the three broad chronological stages (Fig. 6.4).

The comparison based on the taxonomic distinctness index shows that the Palenque faunal record is outside the confidence limits. This occurrence is the product of bias in the recovery strategy as well as in the analysis of the archaeological material (only mammals were recorded and only large bones were retained). Similarly, the Ethnozoological record is slightly out of, but near, the upper confidence interval. We interpret this result to mean that this faunal record is more complete than the archaeological data and could suggest the recent incorporation of new animals to the local uses, as discussed below (Fig. 6.5).

Regardless, we faced the choice of limiting our comparison only to mammals (with an important loss in comparative analysis) or continuing the comparison of all localities while recognizing and emphasizing differences in sample content. In order to develop a preliminary scenario on the historical uses of the vertebrates, we decided to use all of the localities. Nevertheless, we must emphasize that the discussion on mammals may be more reliable than those for birds and herpetofauna.

6. The historical use of terrestrial vertebrates in the Selva Region (Chiapas, México)

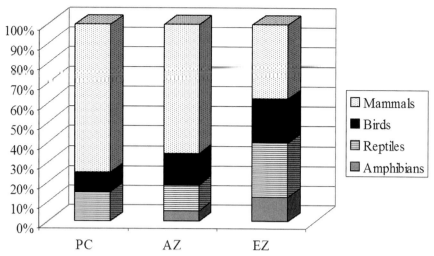

Fig. 6.4 Terrestrial vertebrates grouped by class and chronological period. PC: Pre-ceramic; AZ: Archaeozoological; EZ: Ethnozoological.

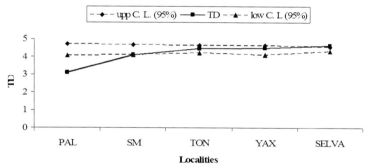

Fig. 6.5 Graphic of the taxonomic distinctness index by locality and upper (upp) and lower (low) 95% confidence limits (C. L.). Pre-Ceramic period: SM, Santa Marta Rock Shelter. Archaeozoological period: PAL, Palenque; TON, Toniná; YAX, Yaxchilán. Ethnozoological period: Selva, Selva region. See text for details.

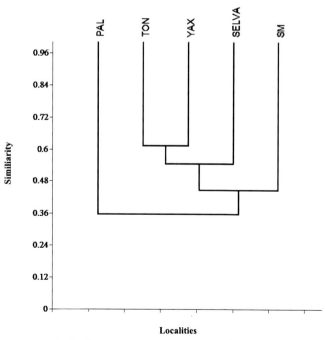

Fig. 6.6 Cluster analysis revealing the similarity of localities. The cluster is based on the Simpson index and UPMGA. Pre-Ceramic period: SM, Santa Marta Rock Shelter. Archaeozoological period: PAL, Palenque; TON, Toniná; YAX, Yaxchilán. Ethnozoological period: Selva, Selva region. See text for details.

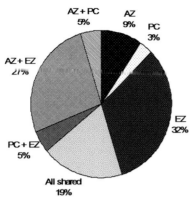

Fig. 6.7 Results of seriation analyses revealing the percentage of taxa exclusive to each chronological stage and that shared by two or all chronological stages. See text for details. PC: Pre-Ceramic; AZ: Archaeozoological; EZ: Ethnozoological.

To corroborate the similarity between localities, a cluster analysis was processed based on the Simpson index, as discussed above (Fig. 6.6). In this case, all archaeological localities with a similarity of above 80% yielded a significant difference with the Ethnozoological record, suggesting diachronic changes in animal use by local peoples in the Selva region.

The results of the unconstrained seriation reveal those taxa recorded in one or more localities. In this case, to facilitate the interpretation, the localities were grouped in the three chronological stages. In Figure 6.7, exclusive and shared taxa are shown as percentages.

Two field rodents (*Baiomys* sp. and *Liomys* sp.) were exclusively recorded in the Pre-Ceramic period. This is interesting, as the deposition of these rodents may have been non-anthropogenic. However, the Ethnozoological record shows that these mammals are used as a food source. Based on this assertion, these results suggest that rodent consumption may have occurred very early with a preference for these species. In later periods, these rodents appear not to have been used, although they may have been substituted by other rodents.

In the Archaeozoological period six exclusive taxa were identified. One is the falcon (*Falco* sp.), identified in the Toniná locality, and three rodents identified only in Palenque (*Neotoma mexicana*, *Tylomis* sp. and *Heteromys* sp.). Additionally, remains of a toad (*Bufo* sp.) and cougar (*Puma concolor*) were both recovered in Yaxchilán and Palenque. While the rodents may be interpreted as within the Archaic period sample, there are no current records of the use of falcon, cougar and toad by local people. This situation might be interpreted as a discontinuation in use of these animals, perhaps due to a change in the cultural attitude toward them. These animals still inhabit the region, and felines and birds of prey were important symbolic animals in Mesoamerica.

Another important result is represented by the 21 taxa that were recorded exclusively within the Ethnozoological period. Nine of these taxa belong to the herpetofauna, eight to birds, and four to mammals. This scenario could be interpreted as the result of recent additions to local faunal use. The mammals were hooded skunk (*Mephitis macroura*), eastern spotted skunk *(Spilogale putorius)*, Mexican porcupine (*Coendu mexicanus*), and northern raccoon (*Procyon lotor*). These animals are part of the local biodiversity and no evidence exists to suggest any change in their geographical range. Also, these animals were well known as a medicinal or food resource in Central México (Barajas Casso 1951; Corona-M 1996). It is probable that this change in cultural attitude toward these animals was produced during the Colonial period, but further historical research is necessary to determine the accuracy of this assertion.

For birds and herpetofauna, one must remember the caveats discussed above. Nevertheless, the occurrence of eight bird taxa is noteworthy. All of these birds are well known for their use in feather artwork by México's people (Corona-M 2002), including parrots (cf. *Amazona* sp.), scarlet macaw (*Ara macao*), resplendent quetzal (*Pharomacrus mocino*), blue-crowned motmot (*Momotus momota*), hummingbirds (Trochiilidae), great egret (*Ardea alba*) and gulls (*Larus* sp.). Others, such as the black vulture (*Coragyps atratus*), parrots, hummingbirds, and egrets, may have been used as a medicinal resource (Corona-M 2002; 2005).

Concerning the herpetofauna, boa (*Boa constrictor*), rattlesnake (*Crotalus durissus*), Mexican burrowing toad (*Rhinophrynus dorsalis*), treefrogs (*Hyla* sp., *Smilisca baudini*), snapping turtle (*Chelydra serpentine*), brown basilisk (*Basiliscus vittatus*), helmeted basilisk (*Corytophanes* sp.) and rainbow ameiva (*Ameiva undulate*) probably represent recent additions to cultural usage, though snakes and turtles were important symbolic animals in Mesoamerica. Nevertheless, for birds and herpetofauna records, further research is necessary to confirm possible recent additions in the uses by local people, especially since pictographic evidence could suggest the ancient use of these animals (Tozzer and Allen 1910; Seler 2004).

Additionally, we identified faunal resources that were used consistently from ancient times through modern times in the Selva region. We considered animals that were present in all chronological stages in both archaeological and ethnozoological records. These animals are represented by three turtle taxa (*Trachemys scripta*, *Dermatemys mawii*, and *Kinosternon* sp.), one species of crocodile (*Crocodylus moreletii*), two species of birds (*Crax rubra* and *Meleagris ocellata*) and mammals including rodents (*Peromyscus* sp., *Ototylomis* sp., *Orthogeomys* sp., *Dasyprocta punctata*, and *Agouti paca*), ungulates (*Tayassu pecari*, *Pecari tajacu*, *Tapirus bairdii*, *Odocoileus virginiana*, and *Mazama americana*), rabbit (*Sylvilagus* sp.), squirrel (*Sciurus* sp.), jaguar (*Panthera onca*), and nine-banded armadillo (*Dasypus novemcinctus*).

The primary use of these animals was for food, but they were also used as raw materials. For example, turtles are currently used to manufacture bags and some musical instruments, while crocodiles and some amphibians are used as local medicines as well as food. However, some

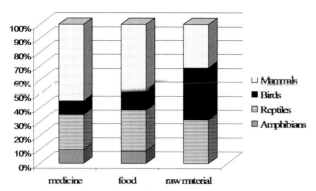

Fig. 6.8 Graphic showing the three broad categories of use by class of terrestrial vertebrates.

of the uses of animal resources cannot be determined, especially in the archaeological record where there is a tendency to bias the interpretation of these faunal remains as being food or raw material or as having symbolic characteristics. Nevertheless, other uses, such as for medicine, cannot be ruled out. In any case, based on the ethnozoological record, possible uses of each animal group are presented in Figure 6.8.

Currently, mammals are used mainly as food, where the rodents and ungulates represent the most important resource and no ungulate is used as medicine. Nevertheless, a strong relationship exists between edible animals and medicinal use, since most of the animals were eaten to prevent some illnesses. As raw materials, mammals are the most important species, particularly the Central American agouti (*Dasyprocta punctata*), jaguar (*Panthera onca*), white tail deer (*Odocoileus virginianus*), and red brocket (*Mazama americana*). Seven birds were used as raw materials, and only the Galliformes were used as food. Only two birds were reportedly used for local medicinal resources: parrots (*Amazona* sp.) and black vulture (*Coragyps atratus*).

Finally, we must point out some differences in the cultural attitudes toward some animals. In spite of the fact that in the Selva region subsistence hunting is a common practice (Guerra 2001), the Lacandon people, who inhabit part of the study region, use animals only for food and as raw materials; no animal is used as a medicinal resource. Other local peoples, such as in the *Montes Azules* area, hunt some animals for specific medicinal resources. This situation illustrates the complexity of reconstructing the historical pattern of faunal utilization.

Conclusion

A conservative comparison of diachronic data reveals, with some reliability, the persistent use of some animals through time in a limited geographic setting. Both sets of data reveal remarkable consistency for the mammalian fauna. There is, however, less consistency between these data sets for other terrestrial vertebrates. This is mainly due to an analytical artefact produced by the typical biases in the archaeological record caused by differential excavation methods and taphonomic process, as well as the intrinsic problems of faunal identification.

The diachronic changes in faunal use provide a less reliable comparison. However, these patterns are offered as initial scenarios that require further research and as a more complete faunal record from archaeozoological sites. In both cases, the persistence and the changes in the use of faunal resources reveal the importance of local resources and the cultural attitudes that govern the consumption of specific animals.

Acknowledgements

We are grateful to William Belcher and Umberto Albarella; their kind comments greatly improved the original manuscript.

References

Álvarez, T. (1976) *Restos óseos de la primera excavación de Santa Martha, Chiapas*. In *Excavaciones en el abrigo de Santa Marta, Chiapas (1974)*. Informes del Departamento de Prehistoria, Instituto Nacional de Antropología e Historia, México.

Álvarez, T. (1982) Análisis del material zoológico de las excavaciones de Toniná, Chiapas. In P. Becquelin and E. Taladire (eds.) *Toniná, un cité Maya du Chiapas (Mexique)*. Tome II. Collection Études Mesoaméricaines 6, 1127–1142. México, D.F., Centre d'Études Mexicaines et Centraméricaines.

Álvarez, T., Ocaña, A. and Valentín, N. (1980). *Identificación de los restos óseos procedentes de las excavaciones de Tonina, Chiapas*. Informe de trabajo. Instituto Nacional de Antropología e Historia, México.

Álvarez, T., Ocaña, A. and Valentín, N. (1990) Identificación de los restos óseos procedentes de las excavaciones de Tonina, Chiapas. In P. Becquelin and E. Taladire (eds.) *Toniná, un cité Maya du Chiapas (Mexique)*. Tome IV. Collection Études Mesoaméricaines 6, 1832–1846. México, D.F., Centre d'Études Mexicaines et Centraméricaines.

Aranda, M. (1997) Comercio de pieles de mamíferos silvestres en Chiapas, México. In J. G. Robinson, K. H. Redford and J. E. Rabinovich (Comp.) *Uso y conservación de la vida silvestre neotropical*, 215–218. México, Fondo de Cultura Económica.

Barajas Casso, L. E. (1951) *Los animales usados en la medicina popular mexicana*. México, Imprenta Universitaria, Universidad Nacional Autónoma de México.

Chavez-O, E. A. (1969) Artifactual and Non-artifactual Material of the Phyla Mollusca, Arthropoda, and Chordata from Chiapa de Corzo, Chiapas. In T. A. Lee (ed.) *The Artifacts of Chiapa de Corzo, Chiapas, Mexico*. Papers of the New World Archaeological Foundation No. 26, 219–220. Provo, Brigham Young University.

Clarke, K. R. and Warwick, R. M. (1998) A taxonomic distinctness index and its statistical properties. *Journal of Applied Ecology* 35, 523–531.

CONABIO (1998) *La diversidad biológica de México: Estudio de País, 1998.* México, Comisión Nacional para el Conocimiento y Uso de la Biodiversidad.

Corona-M, E. (1996) Las aves en el mercado de Sonora. Una prospección etnozoológica. *Vertebrata Mexicana* 2, 3–8.

Corona-M, E. (2002) *Las aves en el siglo XVI novohispano.*

México, Colección Científica, Instituto Nacional de Antropología e Historia.

Corona-M, E. (2005) Archaeozoology and the role of birds in the traditional medicine of Pre-Hispanic Mexico. *Documenta Archaeobiologiae* 3, 293–300.

Corona-M, E. and Arroyo-Cabrales, J. (2003) Las relaciones hombre-fauna, una zona interdisciplinaria de estudio. In E. Corona-M and J. Arroyo-Cabrales (eds.) *Relaciones Hombre-Fauna*, 17–28. México, Plaza y Valdéz-CONACULTA-INAH.

Cupul-Magaña, F. (2003) Cocodrilo: medicina para el alma y el cuerpo. *Revista Biomédica* 14, 45–48.

Emery, K. F. (2004a) Historical Perspectives on Current Research Directions In K. F. Emery (ed.). *Maya Zooarchaeology: New Directions in Method and Theory*. Cotsen Institute of Archaeology at UCLA, Monograph 51, 1–2. Los Angeles, Cotsen Institute of Archaeology.

Emery, K. F. (2004b) In Search of Assemblage Comparability. Methods in Maya Zooarchaeology. In K. F. Emery (ed.) *Maya Zooarchaeology: New Directions in Method and Theory*. Cotsen Institute of Archaeology at UCLA, Monograph 51, 15–34. Los Angeles, Cotsen Institute of Archaeology.

Enríquez, P. (2005) *Uso medicinal de la fauna silvestre en los Altos de Chiapas*. Tesis de maestría. Colegio de la frontera sur, Chiapas, México.

Enríquez, P., Mariaca, R., Retana, O. and Naranjo, E. (2006) Uso medicinal de la fauna silvestre en los Altos de Chiapas. *Interciencias*, 31(7), 491–499.

Flannery, K. V. (1969) An Analysis of Animal Bones from Chiapa de Corzo, Chiapas. In T. A. Lee (ed.) *The Artifacts of Chiapa de Corzo, Chiapas, Mexico*. Papers of the New World Archaeological Foundation No. 26, 209–218. Provo, Brigham Young University.

Flores-Villela, O. (1993) Herpetofauna Mexicana. Lista anotada de las especies de anfibios y reptiles de México, cambios taxonómicos recientes y nuevas especies. *Carnegie Museum of Natural History, Special Publication*. 7, 1–73.

García-Bárcena, J. (1982) *El precerámico de Aguacatenango, Chiapas, México*. Colección científica 110, Serie Prehistoria. México, Instituto Nacional de Antropología e Historia.

García-Bárcena, J. and Santamaría, D. (1982) *La Cueva de Santa Marta Ocozucoautla, Chiapas*. Colección Científica 111, Serie Prehistoria. México, Instituto Nacional de Antropología e Historia.

Gobierno del Estado de Chiapas (2005) *Los Municipios del Estado de Chiapas*. Enciclopedia de los Municipios de México. Instituto Nacional para el Federalismo y el Desarrollo Municipal, Gobierno del Estado de Chiapas. Accessed February 20, 2007, http://www.e-local.gob.mx/work/templates/enciclo/chiapas/.

Góngora-Arones, E. (1987) *Etnozoología lacandona: la herpetofauna de Lacanja-Chansayab*. México, Instituto Nacional de Investigaciones sobre Recursos Bióticos.

Green, D. F. and Lowe, G. W. (1967) *Altamira and Padre Piedra, Early Preclassic Sites in Chiapas Mexico*. Papers of the New World Archaeological Foundation No. 20. Provo, Brigham Young University.

Guerra, M. (2001) *Cacería de subsistencia en dos localidades de la selva lacandona, Chiapas, México*. Tesis de licenciatura. Universidad Nacional Autónoma de México.

Hammer, O., Harper, D. A. T. and Ryan, P. D. (2007) *PAST: Paleontological Statistics, Ver. 1.67 Software Manual*. University of Oslo. Accessed April, 15, 2007, http://folk.uio.no/ohammer/past.

Harris, D. (2006) The Interplay of Ethnographic and Archaeological Knowledge in the Study of Past Human Subsistence in the Tropics. *Journal of the Royal Anthropological Institute* 12(1) 63–78.

Howell, S. N. G. and Webb, S. (1995) *A guide to the birds of Mexico and Central America*. Oxford and New York, Oxford University Press.

MacNeish, R. S. and Peterson, F. A. (1962) *The Santa Marta Rock Shelter, Ocozocoautla, Chiapas, Mexico*. Papers of the New World Archaeological Foundation No. 14. Provo, Brigham Young University.

March, I. J. (1997) Los Lacandones de México y su relación con los mamíferos silvestres: un estudio etnozoológico. *Biótica* 12, 43–56.

Marrinan, R. A. (1986) Appendix: Faunal Analysis. In R. C. Treat (ed.) *Early and Middle Preclassic Sub-Mound Refuse Deposit at Vistahermosa, Chiapas*. Notes on the New World Archaeological Foundation No. 2, 34–37. Provo, Brigham Young University.

Martinez-Muriel, A. (1989) Basureros del Formativo Tardío en Don Martin, Chiapas. *Arqueologia 2a. Epoca* 1, 61–70.

Naranjo, E. J., Rangel, J. L., Vásquez, I. and Hernández, H. G. (1997) *Plan de manejo para la fauna silvestre de la subregión Marques de Comillas, Chiapas*. Chiapas, México, San Cristóbal de Las Casas.

Naranjo, E., Guerra, M., Bodmer, R. and Bolaños, J. (2004) Subsistence hunting by three ethnic groups of the Lacandon forest, Mexico. *Journal of Ethnobiology* 24(2), 233–253.

Ocaña, A. (1997) Estudio de los mamíferos del templo olvidado, Palenque, Chiapas. In J. Arroyo-Cabrales and O. J. Polaco (eds). *Homenaje al profesor Ticul-Álvarez*. Colección científica 357. México, Instituto Nacional de Antropología e Historia.

Ocaña, A. and Polaco, O. J. (1982) *Informe de los restos óseos procedentes de Tonina, Chiapas*. México, Informe del Laboratorio del Arqueozoología, SLAA-INAH.

Ocaña, A and Polaco, O. J. (1990) Informe de restos óseos procedentes de Toniná, Chiapas. In P. Becquelin and E. Taladire (eds.) *Toniná, un cité Maya du Chiapas (Mexique)*. Tome II. Collection Études Mesoaméricaines 6, 1851–1852. México, D.F., Centre d'Études Mexicaines et Centraméricaines.

Reid, F. A. (1997) *A field guide to the mammals of Central America and Southeast of Mexico*. Oxford and New York, Oxford University Press.

Seler, E. 2004. *Las imágenes de animales en los manuscritos mexicanos y mayas*. México, Casa Juan Pablos.

Soto. H. (1998) *Estudio arqueozoológico en la ciudad prehispánica maya de Yaxchilan, Chiapas*. Tesis de licenciatura de la Escuela Nacional de Ciencias Biológicas. Instituto Politécnico Nacional, México.

Toledo, V. M., Alarcón-Chaires, P., Moguel, P., Olivo, M., Cabrera, A. and Rodríguez-Aldabe, A. (2002) Biodiversidad y pueblos indios en México y Centroamérica. *Biodiversitas* 43, 1–8.

Tozzer, A. M. and Allen, G. M. (1910) Animal figures in the Maya Codices. *Papers of the Peabody Museum of American Archaeology and Ethnology, Harvard University* 4(3), 275–372.

Voorhies, B. (1976) *The Chantuto People: An Archaic Period Society of the Chiapas Littoral, Mexico*. Papers of the New World Archaeology Foundation Vol. 41. Provo, Brigham Young University.

Zollá, C. (1994) *Diccionario enciclopédico de la medicina tradicional mexicana*. 2 Tomos. México, Instituto Nacional Indigenista.

7. Pacific Ocean fishing traditions: Subsistence, beliefs, ecology, and households

Jean L. Hudson

Two case studies, one in Polynesia and the other in coastal Peru, are used to examine some of the methodological strengths of ethnozooarchaeological approaches to human ecology in the past. Linguistic evidence in Polynesia suggests some dietary taboos may be linked to family health. Participant-observation with reed boat fishers in Peru documents ecological and social details of family-level subsistence fishing and the potential for surplus. Family-based decisions and associated household-oriented archaeology are argued to provide a useful focus for ethnozooarchaeological research, with analytic value for researchers of diverse theoretical frameworks.

Keywords: fishing, ethnoarchaeology, zooarchaeology, family, household, Polynesia, Peru

Introduction

The fishing traditions of the Pacific Ocean are diverse, as are its peoples. The goal here is not to attempt a comprehensive review of fishing practices along the coastlines of the Pacific Ocean, but rather to select a few examples that illustrate some of the ways that an ethnozooarchaeological approach to fishing can yield useful insights. The focus is on the intersection between fishing practices, subsistence goals, beliefs about fishing, and human ecology. In the process, I will argue for the merits of an analytic focus on the household, rather than the individual, whether theoretically framed as 'genetically selfish' or 'agency-driven.' A second goal is to illustrate the value of a range of ethnozooarchaeological field methods, some more qualitative, others more quantitative.

To do this I will focus on two rather different case studies, one in Polynesia and the other in Peru (Fig. 7.1). In both cases the beliefs and practices of modern fishing people are examined for insights of potential relevance for our understanding of the past.

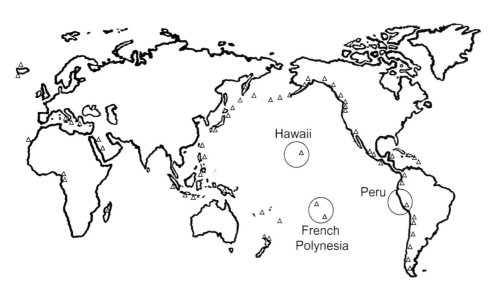

Fig. 7.1 Location of study sites and coastal Holocene volcanism. Triangles approximate locations of volcanoes; source data from the Smithsonian Institution Global Volcanism Program.

Analytic Merits of Focusing on the Household

Before detailing the two cases, I will outline the potential analytic merits, and some of the attendant theoretical issues, in focusing on the household rather than the individual or the society as a whole. I will then look briefly at how archaeologists in the two study areas, Polynesia and Peru, have integrated social and archaeological concepts of household in those regions.

In the North American zooarchaeological literature, there is a strong representation of theoretical approaches that are grounded in evolutionary theory (*e.g.* Broughton 1997; Butler 2000). Many researchers use the quantitative data from their zooarchaeological assemblages to test for evidence of evolutionarily-driven human decisions, with energy optimization as the most common proxy (*e.g.* Byers and Broughton 2004), although alternative currencies, such as prestige, have been suggested (McGuire and Hildebrandt 2005). In evolutionary models, inclusive fitness or relative reproductive success is the ultimate goal, regardless of currency (Smith 2004; see Bird and O'Connell [2006] for a recent review of archaeological applications). The individual becomes the analytic focus, at least at the theoretical level. Typically the focus is on males and their natural-selection-enhancing decisions and behaviors, since variance in reproductive success has the potential to be far more extreme for males than females. Recent exceptions include Bliege Bird's (2007) work on gender and age variations in foraging strategies among the Meriam of the Torres Strait Islands, Australia. Methodologically, however, the zooarchaeological assemblages are often analyzed as a whole. Since these typically represent the accumulated decisions of many individuals, there is some discord between theory and method in the practice of this approach.

At the other end of the theoretical spectrum are models of human behavior that place greater emphasis on conscious social agency, often in the form of post-modern attention to unique and constantly negotiated individual identities and agendas, sometimes with the assumption that concepts of oppression and resistance can be usefully applied to all societies (*e.g.* Brumfiel 1992; Hendon 1996). Here, too, the desired unit of analysis is the individual decision-maker, but the decisions are assumed to be driven by factors other than reproductive success. Individual attributes of social identity, such as gender, class, or membership in a wide variety of other social groupings, are seen as the most significant motivators of human decisions. In practice, at the interpretive stage, individuals are often grouped according to some aspect of their social identity. Then the archaeological residues of opposing memberships, such as male and female, or elite and commoner, are assigned and interpreted.

Both these theoretical frameworks share the same methodological handicap: it is rarely possible to associate an assemblage of zooarchaeological remains with a particular individual and their decisions during life. Consequently, both frameworks might find an analytic focus on the household more methodologically feasible, while still allowing the researcher to integrate data from other social scales, such as individual burials or intra- and inter-site variations, when those data are available. Households do, in some archaeological contexts, leave behind tangible architectural remains or patterns of spatial clustering of artifacts, features, and debris that can be used to delimit likely household groups and their zooarchaeological remains.

Household Archaeology: Theoretical and Methodological Issues

An analytic focus on the house and/or household has a long history in archaeology, one which I will not attempt to review comprehensively here. Notable early works that made productive use of the house to represent a basic social unit within the larger community or settlement include those by Chang (1958) and Flannery (1976). Household archaeology was explicitly advocated, applied and critiqued in the 1980s (*e.g.* Bawden 1982; Kent 1987; Stanish 1989; Wilk and Rathje 1982). More recent comparative works and reviews include Blanton's (1994) cross-cultural comparisons, Hendon's (1996) critique, Allison's (1999) attention to activities within households, and Beck's (2007) applications of the Lévi-Strauss 'house society' concept.

Household archaeology is not without its problems. There is the fundamental issue of how closely our interpretations rest on an exact match between the physical dwellings and the social groupings we assume they represent. We should expect an imperfect match, given all the ways in which residents of one dwelling area can interact with individuals located elsewhere in the community and larger region, but we might reasonably hope that the match will be sufficient for tackling questions about often-repeated behaviors that leave archaeologically robust material patterns. We should expect variability; there is no need to assume or decry a normative approach. We should expect that most archaeological accumulations will represent a mix of decisions and decision-makers. Comparisons among households within communities and across regions will be needed to discern both commonalities and variations. We should be attentive to the match between the time span of archaeological accumulation – a season, a few years, several generations – and the particular questions we seek to answer about the past.

There are also practical issues. Preservational filters may allow or blur the definition of separate dwellings, complexes or compounds and their associated middens. People with durable dwellings will be easier to study than those with ephemeral ones; a household approach will not be equally feasible for all times and places. Excavation and analysis costs may limit the number of contemporaneous dwellings available for study and require special budgeting at the proposal stage of research.

It is, however, possible to design and conduct meaningful research, given these issues of thoughtfully matching methods and questions. Household-oriented archaeology has been productively pursued in both the study areas of concern here, Polynesia and Peru.

Household Studies in Polynesia and Peru

In their recent review of Polynesian archaeology, Kirch and Kahn (2007) note the application by themselves and others working in Oceania of the Lévi-Strauss concept of house societies. This concept focuses attention on the social and ideological role of house and household in creating identity and economic relationships, especially in societies where blood lineages are less important (see Beck [2007] and Gillespie [2007] for thoughtful discussion of the original "*sociétés à maisons*" concept and its archaeological applications). Kahn (2007) applies a household approach to her examination of status distinctions between different dwelling areas within a larger residential and ceremonial complex on the island of Moorea. O'Day (2004) and Kirch and O'Day (2003) explore the integration of ethnography and archaeology and some of the zooarchaeological evidence for status differences and female roles in Polynesian fishing societies.

Archaeologists working in Peru and other parts of Andean South America have also made productive use of household archaeology, using a diverse array of framing theories. Among the many valuable contributions that could be cited are Aldenderfer's (1988) work at the Archaic site of Asana and his (1993) edited volume on domestic architecture and ethnicity, Bawden's (1982) residential comparisons at Galindo, Chapdelaine's (2002) detailing of Moche urban life, Isbell's (1996) discussion of Andean social concepts as they relate to archaeological remains of households, Janusek's (2004) work at Tiwanaku, Stanish's (1989) discussion of household archaeology as an analytic method for approaching ethnicity, and Vaughn's (2005) analysis of Nasca domestic architecture. Specific attention to zooarchaeological data at the household level is not yet common, but published examples, such as the work by Roselló, Vásquez, Morales and Rosales (2001) in an urban Moche sector, hint at the potential.

It thus seems worth taking a closer look at how a household level analysis of zooarchaeological and ethnozooarchaeological data might help us bridge a common gap between theory and method, the gap between our theoretical models of individual decision makers (however they might be motivated) and our recovery of socially and temporally aggregated archaeological remains. The two case studies that follow illustrate some of the ways that diverse ethnozooarchaeological methods might add to our insights about human–animal relationships in the past, and how a household scale approach to our faunal data might be relevant and useful.

A Polynesian Case Study

The Polynesian case study comes from an interdisciplinary project begun in 2005 in French Polynesia (Fig. 7.2), specifically on the islands of Tahiti, Moorea, and Tikehau, a project whose premise is outlined in the journal Clinical Toxicology (Dellinger *et al.* 2005). The project grew out of a conversation between a Polynesian translator, Hinano Murphy, who is also president of a Tahitian heritage group, *Te pū ātitiā*, and a toxicologist, John Dellinger, from the University of Wisconsin, Milwaukee. Hinano was explaining a Tahitian dietary taboo that specifies that when you are pregnant or nursing, you cannot eat "outside" or *e-i'a tua* fish. Outside fish refer to those found outside the lagoon, such as pelagic tuna. These outside fish include piscivorous species and older individuals. Ecologically, it is these outside fish that concentrate toxic levels of mercury, or more precisely, methylmercury. From the perspective of human consumers of such fish, the family members most vulnerable to mercury poisoning are developing infants and children (Myers and Davidson 1998). This dietary taboo thus does a good job of protecting the health of women and children.

I was invited to join the project to add an anthropological and archaeological perspective. I found good linguistic evidence that a very similar taboo was part of traditional Hawaiian beliefs. Titcomb (1972, 17–18), citing Cobb (1900/1901), Pogue (1858), and Malo (1903), notes that on the Hawaiian islands certain fish were taboo for pregnant women, specifically *aku* (ocean bonito or skipjack tuna, *Katsuwonus pelamis*) and *'opelu* (mackerel scad, *Decapterus* sp.). These are both predatory fish of the open ocean, ecologically positioned to accumulate toxic levels of methylmercury. Hawaiians and Tahitians share Polynesian heritage (Kirch 2000), but they are separated by enough water and linguistic variation for the similarity

Fig. 7.2 Polynesian case study. As part of the 2005 fieldwork in French Polynesia, locally caught fish, such as this Tatihi (Naso brevirostris), were sampled for their mercury content.

of their dietary taboos to suggest significant antiquity and cultural importance.

This is where the research of a fourth member of the team, David Krabbenhoft, a geologist, made a significant addition. Volcanoes are natural sources of atmospheric mercury. Krabbenhoft and Schuster (2002) used ice core data to track historical patterns of atmospheric mercury. Their work demonstrates that while our modern industrial practices are responsible for the steady level of methylmercury accumulation in fish that we worry about today, some of the largest historic volcanic eruptions, such as that of Krakatoa in 1883, created brief atmospheric mercury spikes even higher than those we produce industrially. The Pacific Ocean is of course geographically surrounded by active volcanic systems, and some of the Polynesian islands are themselves volcanic (Fig. 7.1).

This means that it is quite conceivable that ancient Polynesians experienced short bursts of mercury toxicity in the fish they relied upon for so much of their subsistence. The impacts of mercury on developing infants is dramatic and visible, resulting in a variety of birth defects including microcephaly, cerebral palsy, seizures, and mental retardation (Myers and Davidson 1998). The stage was thus set for Pacific Ocean people to observe correlations between the diets of men and women, and among various pregnant women, and discern the negative impacts when pregnant women ate outside fish.

Dietary taboos represent another popular anthropological topic of study to which I cannot hope to do justice in a paper of this scale. I am defining taboos as conscious, verbally articulated social rules about appropriate behaviour, in this case rules about what you should or should not eat. Archaeologically, we have little access to what was verbally articulated in the past. Thus to frame possible explanations of taboos anthropologists rely heavily on ethnographic and linguistic observations. Once hypotheses are framed, archaeological data can contribute to tests of their relevance based on the presence or absence of expected material results, such as differences in diet. Ethnozooarchaeology is uniquely suited to tackling such issues, as it explicitly engages both ethnographic and archaeological types of data.

Perhaps of most immediate relevance to the themes discussed here are discussions of food-related taboos within Oceania and circum-Pacific regions. Some have focused more on inter-household differences in status (Kahn 2007; Kirch and O'Day 2003); dietary restrictions that differentiate men and women are mentioned, but chiefly vs. commoner differences are the primary focus. Others have highlighted links between beliefs and ecological sustainability (Swezey and Heizer 1977; Watanabe 1973) or among beliefs, nutrition and health, seasonal use of specific fish habitats, and sustainability (Rouja, 1998; Rouja *et al.* 2003). Another significant body of literature on fishing peoples of Oceania focuses on gender differences in food acquisition rather than food consumption, some theoretically framed by symbolism (Brightman 1996), others by evolutionary ecology (*e.g.* Bliege Bird *et al.* 2001;

Sosis 2000). The evolutionary models for prey choice, costly signalling, and symbolic capital provide interesting hypotheses for how and why particular foods are obtained, and for division of labor and patterns of food sharing, but do not focus directly on individual or gender differences in foods consumed and their impacts on health.

The correlation of dietary taboos based on gender and the resulting health of children is a working hypothesis to explain this Polynesian dietary taboo. What makes it especially relevant to the issues raised earlier in the discussion of theoretical frameworks is that what might otherwise have been judged a self-serving monopoly by men on the biggest and the best of the fish becomes open to a very different interpretation. Adult men can eat mercury-rich tuna with far less risk to their health. This division of diet by sex and age thus serves the best interests of many members of the family. It provides an interesting alternative to male-oriented or individual-oriented explanations, shifting the focus to shared interests at the family or household level. At the same time, it allows a theoretical fit with both evolutionary models and agency models, since family health can be conceived of as a goal of both inclusive fitness and socially motivated agendas.

Two other aspects of this case study are of interest as well. First, it is potentially testable through several lines of evidence. Such evidence includes: dated archaeological contexts corresponding to major volcanic events visible stratigraphically; changing patterns in zooarchaeological fish remains; mercury levels in associated human remains; and a more thorough examination of relevant linguistics. Secondly, the pairing of volcanic activity and dietary reliance on fish is not unique to Polynesia. Productive coastal fisheries and Holocene volcanism co-occur in many parts of the world (Fig. 7.1), including the Caribbean, the Mediterranean, and much of the Pacific coast of Asia and the Americas, many of which have archaeological records of human reliance on fishing. Future research could explore the relevance of gender-related family dietary patterns and fish toxicity in such contexts, and test the wider relevance of health-seeking goals in general and family health-related decisions in particular as important aspects of past human relationships within the ecology of fishing.

A Peruvian Case Study

The second case study takes us to a different part of the Pacific Ocean, the west coast of Peru. Since 2001 I have been conducting ethnographic work with modern fishing families in a Peruvian coastal community. I chose this particular community because of their reliance on reed boats as watercraft when fishing the ocean. So many modern fishing communities throughout the world have adopted motorized craft that this seemed a rare and valuable opportunity to collect information about non-motorized styles of fishing, styles that might compare in a meaningful way with those of the archaeological past.

Ethnographic analogy must be done thoughtfully, as no modern case will be identical in all parameters to any archaeological case. Ideally one recognizes points of divergence as well as similarity, seeks meaningful parallels in ecology, economy, and social and political life, and considers carefully how long the studied modern population has been in its current ecological setting and what particular historic trajectories of change have been at work. One also seeks to disentangle underlying relationships of cause and effect, and assess their relevance to the particular archaeological case of interest. For example, if modern Peruvian reed boat fishing usually involves division of labor within an extended family, what needs are met by this particular social approach and what alternatives might also serve those needs? How relevant are those needs for fishing communities of the past? In this particular case, the benefits of certain patterns of division of labor can be linked to parameters with long-term relevance, such as the ecological behaviours of different fish species, the physical demands of paddling through rough surf, and the efficiency and security of pooling labor or resources.

In this part of Peru we also have the benefit of a thread of continuity stretching from the present back through recent historic records (Edwards 1965), colonial period ethnohistoric accounts (*e.g.* Garcilaso de la Vega [1609] as translated by Livermore [1966]; Rostworowski de Diez Canseco 1981), and into the more distant archaeological past. Reed boats are illustrated in Moche and Chimu art from the first millennium AD (Benson 1972; Donnan 1978; McClelland 1990). Archaeological remains of bundled fishing nets with their stone weights and gourd floats still attached have been recovered from strata radiocarbon dated to roughly 4000 years ago; the site of these finds, Huaca Prieta, is located less than 100km to the north of the modern fishing community discussed here (Bird and Hyslop 1985; Hudson 2004). In this case, then, we may argue that many parameters of the ethnographic analogy are well matched. Care is taken in the discussion that follows to consider aspects of human behavior and decision-making that are likely to have deep temporal relevance, and to use ethnography as a source of insight, rather than to paste the historically impacted particulars of a modern case onto the past.

I began the Peruvian study with ecological questions. Which species of fish were being caught from reed boats and which from shore? How productive was the fishing? How many hours did they spend at it? How many kilograms did they catch? How did particular fishing techniques affect the amount and kind of fish caught? Was fishing with nets rather than hooks significantly different in terms of either absolute productivity or relative gain for labor invested? These questions are relevant ones for those of us who try to interpret zooarchaeological assemblages of fish remains. In Andean archaeology, the ability of coastal fisheries to provide a large enough surplus to set in motion the emergence of political complexity has been long discussed, as has the significance of the use of both nets and watercraft in accumulating such a surplus (*e.g.* Moseley 1975; 1992; Haas and Creamer 2006).

Results of the ethnographic work thus far have been useful. Based on data collected during three different field seasons (2001, 2003 and 2004), and including both seasonal and annual variations, I have some quantitative answers to those questions (Table 7.1). The catch per fishing event can vary between 5kg and 60kg (N=65 fishing events). Fishing events can be as brief as half an hour or as long as eight hours. Several events can occur in a single day, and the accumulation of surplus beyond subsistence needs is a common occurrence. In other words, the combined daily catch of the household often exceeds what is needed for all its members' meals by a significant margin. For example, an extended family with 10 members could accrue 100kg of seafood in a day; if each member consumed a kilogram, they would still be left with a 90kg surplus.

Median catch size for hook and line versus nets (both as practiced from reed boats) is actually very similar, 13 to 16kg, respectively. However, the largest catches, when they do occur, tend to be netted from boats. Also interesting is that the taxonomic families of fish caught by each technique show considerable overlap; for example, fish taxa caught with nets from watercraft may also be caught with nets from shore, and fish taxa caught with hooks may also be caught with nets. Furthermore, the reality of daily fishing within any particular household mixes techniques, and household middens accumulated over any period of time would also reflect a mixture.

Methods and Insights

This style of ethnozooarchaeology – quantitative time studies and enumeration of catches per fishing event, based on participant-observation – can be very productive. As a zooarchaeologist I learned that even in modern times, when the Peruvian fisheries have been greatly reduced by international commerce, a reed boat fisher could gather quite a bit of daily surplus. This suggests that ecologically the Maritime Foundations hypothesis (*sensu* Moseley 1975; 1992) for the rise of complex regional polities along the North Coast of Peru is on firm footing. This hypothesis argues that highly productive fisheries can, like agriculture, create food surpluses, support large sedentary communities, foster specialization, and underwrite new social experiments, including large-scale cooperative activities, such as monument building or competitive social hierarchies. Prehistorically, Peruvian fishing communities could certainly have caught much more than they needed for family level subsistence. The impacts of historic changes work against surplus accumulation by modern reed boat fishers. These historic impacts include decimation of the fisheries themselves by industrialized fishing, fewer family members devoting their working time to fishing activities while still consuming the catch, and decreased use of shellfish and marine plants. Yet families continue to accumulate enough fishing surplus to make a viable livelihood.

I also learned that in a single day the catch could come from multiple habitats: surf zone, beyond the surf zone, sandy bottom, rocky bottom, and at various depths in

	Fishing from Shore	Fishing from Watercraft	
	With Nets	With Nets	With Hooks
Number of fishers needed	1 to 3 people	1 to 2 people	1 person
Hours per fishing event	0.5 to 1.5 hours	0.75 to 8 hours	1 to 6 hours
Kilograms per fishing event 1) median 2) range 3) N (events observed)	under 5kg 1 to 10kg N=16	16kg 5 to 60kg N=19	13kg 5 to 35kg N=30
Fish families represented: Ariidae Atherinopsidae Blennidae Carangidae Cheilodactylidae Clupeidae Engraulidae Haemulidae Labrisomidae Merluccidae Mugilidae Myliobatidae Paralichthydae Rhinobatidae Sciaenidae Scombridae Squatinidae Triakidae	yes yes yes yes yes yes yes yes yes yes	yes yes yes yes yes yes yes yes yes yes yes	 yes yes yes yes yes yes yes yes yes yes

Table 7.1 Quantitative results from ethnozooarchaeological research in Peru.

the water column. This provided a gentle reminder that archaeologists should avoid overly simplistic assumptions that match a particular archaeological assemblage of fish remains with a single type of prehistoric fishing in terms of ecological niche exploited or technologies utilized.

However, after spending weeks helping to pull in nets at dawn, watching and recording the coming and going of reed boaters, and counting the numbers and types of fish caught in a given fishing event, I also became aware of other vitally important aspects of fishing. These were social aspects and they were complex in their patterns of interdependence. I came to realize that the social group that mattered most in this fishing community was the extended family. I am using the term extended family here to contrast with that of nuclear family. If a nuclear family represents one married couple and their dependent children, then an extended family incorporates more than one such set, typically related by some form of kinship. Some Andeanists prefer the term multi-nuclear family for this social arrangement (Isbell 1996).

How might such important social aspects of fishing be studied archaeologically? When possible, analysis of household-level assemblages can provide meaningful subsets of the larger site assemblage. The degree to which those sub-assemblages vary can provide insights into the social composition of the community and open relevant analytic doors to researchers of differing theoretical orientations, regardless of their position within the evolutionary-agency spectrum. For example, prehistoric styles of political integration, be they egalitarian balancing of near-equals, pyramids of ranked differences, communities of specialists provisioning urban centers elsewhere, or some other form, can be meaningfully pursued with household data. Changes over time in household subsistence, division of labor, pooling of resources, and integration within the larger community and region can be studied, as can degrees of variation among households. This is a common approach with other forms of archaeological data, but somewhat neglected in zooarchaeology.

The following narrative is a composite of typical daily activities, some of which are illustrated in Figure 7.3. It demonstrates the possible roles of various family members and indicates why the household would be a productive focus of analysis for both zooarchaeological research and for research into human ecology in general.

The day starts before dawn. An older man – the grandfather in the family – hikes down the coast to the sandy stretch of beach where he keeps his net tied off. The net is anchored on land and floats in the surf zone, catching fish while the family sleeps. He starts the job of pulling the net in to shore. This requires considerable muscle and he is soon joined by his son-in-law and a teen-aged grandson.

Fig 7.3 Peruvian case study. Division of labor within the extended family. Women and children meet incoming reed boat fishermen on the beach to process and market the catch.

They haul in the net and strip it of fish, then let the net back out into the sea. It has been a fair morning and they are bringing a dozen fish home, a total catch of over 5kg, some of which will be had for breakfast, and some of which will be taken by his daughter to market to sell or exchange for other needed items. While the shore fishers are walking home, two adult sons have launched their reed boats to check gill nets set with floats and weights in deeper near-shore waters beyond the surf zone, paddling out and back in an hour and bringing home another 20kg of fish. As the morning wears on, the grandfather walks down to the shore to mend nets while the sons and grandson head out in their reed boats to hook and line a school of fish passing through the bay. The fish are biting and each of the three men brings in another 10 to 15kg. The adult daughter meanwhile is trading some of the surplus fish for crab bait, knowing the grandson will go out crabbing in the afternoon. The midday heat brings everyone in for a meal and a rest. In the afternoon the two sons paddle out to check their gill nets again, the grandfather builds a new reed boat for his grandson – a job of two hours – and then heads down the beach with the son-in-law to do a late afternoon check of his beach net. The three pre-teen children in the family may join their mother at the beach to meet the reed-boaters and help sort the fish. The combined afternoon fishing adds another 25kg to the family catch. The grandson, back from crabbing, contributes part of his 30kg to dinner and converts some of the remainder into a late night of *chicha* (maize beer) with his friends. His grandfather gives him grief the next morning when he shows up late and a bit hung-over for the early beach net pull. But the family – a grandfather, two adult sons, a son-in-law, an adult daughter, three small children, and a teen-aged grandson – has brought in over 100kg of seafood in a single day – a surplus well beyond the needs of family subsistence.

This rich narrative approach is another style of ethno-zooarchaeology, different from and complementary to the quantitative style described previously. One of its benefits is the reminder it provides of the elaborate detail of human social lives, a level of detail we cannot see archaeologically, but whose summed results we do recover. It reminds us to consider the family, and its archaeological manifestation, the house and its midden, as a meaningful unit of zooarchaeological analysis. In this case an extended family of three generations split fishing tasks and associated marketing of surplus and maintenance of fishing gear, and shared the results. The young adult men did more of the physically demanding labor, the older man did more of the work requiring slow-paced discipline and technical expertise, and the woman managed the resulting catch, including the marketing of surplus and handling of family subsistence. In a single day this extended family mixed nets and hooks, shore and boats, subsistence and surplus.

As with the Polynesian case, the door is open to put these ethnozooarchaeological insights to archaeological use. Household-oriented zooarchaeology might shed light on the intersection of social and economic aspects of fishing along the Peruvian coast in the past. Hypotheses about fishing as a household or community specialization within the context of emerging regional political complexity could be tested, as could the chronological depth of household level autonomy and extended family division of labor.

Conclusions

How do these two case studies, one in Polynesia and the other in Peru, contribute to the original goals of this paper? How do they illustrate both ethnozooarchaeological methods and some of the specific ways that fishing practices, subsistence goals, and human social values and beliefs intersect as part of human ecology?

One goal of this paper was to illustrate the value of a range of ethnozooarchaeological field methods, some more qualitative, others more quantitative. In the Polynesian case linguistic data was key in forming a new and archaeologically testable hypothesis about fishing, diet, and family health. In the Peruvian case quantitative time and capture studies were combined with the kind of detailed narrative that participant observation allows. In both cases the ethnographic or linguistic data were used primarily to provide insights relevant for building testable hypotheses about the past, thereby contributing to zooarchaeological research design for future projects.

Both case studies viewed human ecology at the level of the family or household, recognizing that both those concepts, family and household, contain a great deal of cultural variability. Ethnoarchaeological research allows one to see a social group, such as a family, in action, and be persuaded of its relevance to understanding life in the past as well as the present. It also makes real the complexities of family life, including the variations within and between communities in what constitutes a family and the inevitably dynamic internal quality of a family as different members are born, grow, marry, age, and die. Yet for all its variance, it remains an important emic reality, combining special qualities of shared economy and social identity.

Equally important to an archaeologist, household-oriented research is methodologically feasible in ways that individual-oriented research often is not. The record we recover typically represents an accumulation of materials over many days, or seasons, or years, as well as the blending of the results of decisions of many individuals, including males and females of various ages. A household-oriented approach to zooarchaeological analyses could improve our ability to see variance within and between larger social groupings, as well as track changes over time at a scale of decision-making that is both socially meaningful and archaeologically visible.

A second goal of this paper was to illustrate how ethnozooarchaeology, as an integration of ethnography and zooarchaeology, can yield useful insights and alternative models for past human behaviours. Two case studies, one in Polynesia and one in coastal Peru, looked at examples of how fishing practices can combine family-based goals for nutrition, health, and surplus, and can link values and beliefs with subsistence and ecology at the household level. In the Polynesian case a testable model was outlined to explain dietary taboos that protect pregnant women and developing children from eating potentially toxic fish. This provides an alternative to currently popular models focused on male prestige and chiefly prerogatives. In the Peruvian case, the division of labor within families was described and quantitative data on fishing productivity and surplus was presented. The implications of these data for understanding the ecology, economy, and emerging political complexity within prehistoric Peruvian fishing communities were highlighted. In the process, the methodological merits of a theoretical and analytic focus on the household, rather than the 'genetically selfish' or 'agency-driven' individual, were argued.

Acknowledgments

My warm thanks to the fishing families that made the fieldwork reported here possible. The research travel would not have happened without the funding support of the University of Wisconsin and its Center for Latin American and Caribbean Studies in Milwaukee. Many thanks also go to Umberto Albarella for organizing and editing this collection of ethnozooarchaeological studies, and to an anonymous reviewer for helpful suggestions. The Whiteley Center at the University of Washington Friday Harbor Labs provided invaluable support during writing.

References

Aldenderfer, M. S. (1988) Middle Archaic period domestic architecture from southern Peru. *Science* 241(4874), 1828–1830.

Aldenderfer, M. S. (1993) *Domestic Architecture, Ethnicity, and Complementarity in the South-central Andes*. Iowa City, University of Iowa Press.

Allison, P. M. (1999) Introduction. In P. M. Allison (ed.) *The Archaeology of Household Activities*, 1–18. London, Routledge.

Bawden, G. (1982) Community organization reflected by the household: a study of Pre-Columbian social dynamics. *Journal of Field Archaeology* 9(2), 165–181.

Beck, R. A. (2007) The Durable House: Material, Metaphor, and Structure. In R. A. Beck (ed.) *The Durable House: House Society Models in Archaeology*. Center for Archaeological Investigations Occasional Paper 35. Carbondale, Southern Illinois University.

Benson, E. (1972) *The Mochica, A Culture of Peru*. New York, Praeger Publishers.

Bird, D. W. and O'Connell, J. F. (2006) Behavioral ecology and archaeology. *Journal of Archaeological Research* 14, 143–188.

Bird, J. B. and Hyslop, J. (1985) The Preceramic excavations at the Huaca Prieta, Chicama Valley, Peru. *Anthropological Papers of the American Museum of Natural History* 62(1), 1–294.

Blanton, R. E. (1994) *Houses and Households: A Comparative Study*. New York, Plenum Press.

Bliege Bird, R. (2007) Fishing and the sexual division of labor among the Meriam. *American Anthropologist* 109(3), 442–451.

Bliege Bird, R., Smith, E. A. and Bird D. W. (2001) The hunting handicap: costly signalling in male foraging strategies. *Behavioral Ecology and Sociobiology* 50, 9–19.

Brightman, R. (1996) The sexual division of foraging labor: biology, taboo, and gender politics. *Comparative Studies in Society and History* 38(4), 687–729.

Broughton, J. (1997) Widening diet breadth, declining foraging efficiency, and prehistoric harvest pressure. *Antiquity* 97(274), 845–862.

Brumfiel, E. M. (1992) Distinguished Lecture in Archaeology: Breaking and Entering the Ecosystem – Gender, Class, and Faction Steal the Show. *American Anthropologist* 94(3), 551–567.

Butler, V. (2000) Resource depression on the Northwest Coast of North America. *Antiquity* 74(285), 649–661.

Byers, D. and Broughton, J. (2004) Holocene environmental change, artiodactyl abundances, and human hunting strategies in the Great Basin. *American Antiquity* 69(2), 235–255.

Chang, K. C. (1958) Study of the Neolithic social grouping: examples from the New World. *American Anthropologist* 60(2), 298–334.

Chapdelaine, C. (2002) Out in the streets of Moche: urbanism and socio-political organization at a Moche IV urban center. In W. H. Isbell and H. Silverman (eds.) *Andean Archaeology I: Variations in Sociopolitical Organization* 53–88. New York, Kluwer Academic and Plenum Publishers.

Dellinger, J., Hudson, J., Krabbenhoft, D. and Murphy, H. (2005) Pacific volcanoes, mercury contaminated fish, and polynesian taboos. *Clinical Toxicology* 43(6), 595–596.

Donnan, C. (1978) *Moche Art of Peru, Pre-Columbian Symbolic Communication*. Museum of Cultural History, University of California, Los Angeles.

Edwards, C. (1965) *Aboriginal Watercraft on the Pacific Coast of South America*. Berkeley, University of California Press.

Flannery, K. V. (1976) *The Early Mesoamerican Village*. New York, Academic Press.

Garcilaso de la Vega (1609) *Royal Commentaries of the Incas, Part One*. Translated from Spanish by H. Livermore (1966). Austin, University of Texas Press.

Gilliespie, S. D. (2007) When is a house? In R. A. Beck (ed.) *The Durable House: House Society Models in Archaeology*. Center for Archaeological Investigations Occasional Paper 35. Carbondate, Southern Illinois University.

Haas, J. and Creamer, W. (2006) Crucible of Andean civilization, the Peruvian coast from 3000–1800 BC. *Current Anthropology* 47(5), 745–775.

Hendon, J. A. (1996) Archaeological approaches to the organization of domestic labor: household practice and domestic relations. *Annual Review of Anthropology* 25, 15–61.

Hudson, J. L. (2004) Additional evidence for gourd floats on fishing nets. *American Antiquity* 69(3), 586–587.

Isbell, W. H. (1996) Household and *ayni* in the Andean past. *Journal of the Steward Anthropological Society* 24(1 and 2), 249–295.

Janusek, J. W. (2004) Household and city in Tiwanaku. In H. Silverman (ed.) *Andean Archaeology*. Malden, Blackwell Publishing.

Kahn, J. G. (2007) Power and precedence in ancient house societies: a case study from the Society Island chiefdoms. In R. A. Beck (ed.) *The Durable House: House Society Models in Archaeology*. Center for Archaeological Investigations Occasional Paper 35. Carbondale, Southern Illinois University.

Kent, S. (1987) *Method and Theory for Activity Area Research: An Ethnoarchaeological Approach*. New York, Columbia University Press.

Kirch, P. V. (2000) *On the Road of the Winds, An Archaeological History of the Pacific Islands before European Contact*. Berkeley, University of California Press.

Kirch, P. V. and Kahn, J. G. (2007) Advances in Polynesian prehistory: a review and assessment of the past decade (1993–2004). *Journal of Archaeological Research* 15, 191–238.

Kirch, P. V. and O'Day, S. J. (2003) New archaeological insights into food and status: a case study from pre-contact Hawaii. *World Archaeology* 34(3), 484–497.

Krabbenhoft, D. and Schuster, P. (2002) Glacial ice cores reveal a record of natural and anthropogenic atmospheric mercury deposition for the last 270 years. *U. S. Geological Survey, Fact Sheet FS-051-02*, June.

McClelland, D. (1990) A maritime passage from Moche to Chimu. In M. Rostworowski de Diez Canseco and M. Moseley (eds.) *The Northern Dynasties Kingship and Statecraft in Chimor*, 75–106. Washington, D.C., Dumbarton Oaks Research Library and Collection.

McGuire, K. and Hildebrandt, W. (2005) Re-thinking Great Basin foragers: prestige hunting and costly signaling during the Middle Archaic period. *American Antiquity* 70(4), 695–712.

Moseley, M. (1975) *The Maritime Foundations of Andean Civilization*. Menlo Park, Cummings Publishing Company.

Moseley, M. (1992) Maritime foundations and multilinear evolution: retrospect and prospect. *Andean Past* 3, 5–42.

Myers, G. and Davidson, P. (1998) Prenatal methylmercury exposure and children: neurologic, developmental, and behavioral research. *Environmental Health Perspectives* 106(3), 841–847.

O'Day, S. J. (2004) Past and present perspectives on secular ritual: food and the fisherwomen of the Lau Islands, Fiji. In S. J. O'Day, W. Van Neer and A. Ervynck (eds.) *Behaviour Behind Bones: The Zooarchaeology of Ritual, Religion, Status, and Identity*, 153–161. Oxford, Oxbow Books.

Roselló, E., Vásquez, V., Morales, A. and Rosales, T. (2001) Marine resources from an urban Moche (470–600 AD) area in the 'Huacas del Sol y de la Luna' archaeological complex (Trujillo, Peru). *International Journal of Osteoarchaeology* 11, 72–87.

Rostworowski de Diez Canseco, María (1981) *Recursos Naturales Renovables y Pesca, Siglos XVI y XVII*. Lima, Instituto de Estudios Peruanos.

Rouja, P. M. (1998) *Fishing for Culture: Toward an Aboriginal Theory of Marine Resource Use among the Bardi Aborigines of One Arm Point, Western Australia*. Unpublished Ph.D. dissertation. University of Durham.

Rouja, P. M., Dewailly, E., Blanchet, C. and the Bardi Community (2003) Fat, fishing patterns, and health among the Bardi people of North Western Australia. *Lipids* 38(4), 399–405.

Smith, E. (2004) Why do good hunters have higher reproductive success? *Human Nature* 15(4), 343–364.

Sosis, R. (2000) Costly signalling and torch fishing on Ifaluk Atoll. *Evolution and Human Behavior* 21, 223–244.

Stanish, C. (1989) Household archaeology: testing models of zonal complementarity in the south central Andes. *American Anthropologist* 91(1), 7–24.

Swezey, S. L. and Heizer, R. F. (1977) Ritual management of salmonid fish resources in California. *Journal of California Anthropology* 4(1), 6–29.

Titcomb, M. (1972) *Native Use of Fish in Hawaii*. With the collaboration of Mary Kawena Pukui. Honolulu, University of Hawaii Press.

Vaughn, K. V. (2005) Household approaches to ethnicity on the south coast of Peru: the domestic architecture of early Nasca society. In R. Reycraft (ed.) *Us and Them: The Assignation of Ethnicity in the Andean Region, Methodological Approaches*, 86–103. Los Angeles, University of California Los Angeles Institute of Archaeology Press.

Watanabe, H. (1973) *Ainu Ecosystem: Environment and Group Structure*. American Ethnological Society Monograph 54. Seattle, University of Washington Press.

Wilk, R. R. and W. L. Rathje (1982) Household archaeology. *The American Behavioral Scientist* 25(6), 617–639.

8. The ethnography of fishing in Scotland and its contribution to icthyoarchaeological analysis in this region

Ruby N. Cerón-Carrasco

Primary data for archaeological analysis consists of the material remains of past cultures within their spatial and environmental contexts. However, most archaeological analysis relies on the use of additional sources to interpret this primary data. Studies of human history as well as actualistic research extend this potential source.

Actualistic studies include ethnographic parallels and ethnohistory which are most frequently used in the reconstruction of past social behaviour. In Scotland, traditional ethnographic haunts have provided a wealth of information that have allowed construction of analogies to aid the interpretation of archaeological fish bone remains recovered in this region. This work aims to illustrate such ethnographic patterns.

Keywords: actualistic studies, ethnography, fishing, Scotland

Introduction

In most areas in the world, cultures dependant on fishing, hunting and gathering preceded those based on husbandry and/or farming, and this was certainly the case in Scotland. For thousands of years people in the coastal areas of Scotland engaged in various forms of fishing as their primary means of subsistence, beginning with the first recognised human occupation dating to the Mesolithic period *c.* 10,000 years BP. The western seaboard is one of the regions where archaeological sites dating to the Mesolithic have produced large amounts of mainly molluscan remains as well as substantial amounts of fish remains, *e.g.* at Oronsay (Mellars 1978; Mellars and Wilkinson 1980), Carding Mill Bay (Conock *et al.* 1992), Sand (Inner Sound of Skye) (Hardy and Wickham-Jones 2002; 2009), and Port Lobh (Skye) (Finlay forthcoming). Thus, from the earliest of human occupation, exploitation of marine resources began mainly on the seashore where hunter-gatherers found abundant shellfish food; not only were shellfish such as limpet (*Patella vulgata*), mussel (*Mytilus edule*), oyster (*Ostrea ostrea*), winkle (*Littorina littorea*) and cockle (*Ceratoderma edule*) abundant, they were a reliable year-round resource. However, knowledge of the potential of fish as a superior food resource would soon be recognised and human ingenuity give rise to the manufacture of gear for fishing, such as spears, nets, hooks, fishing lines, *etc.* Fishing was first a subsistence occupation, one of the many in the lives of coast dwellers throughout Scotland. It would remain so for most people in Scotland until at least the 15th century AD (Coull 1996) even as the sea was filling with the boats of those who were forced to adopt fishing as their economic mainstay *i.e.* commercial fishing.

In our quest for evidence of fishing in the prehistory of Scotland, we would therefore expect to find some form of fish-catching equipment such as boats, lines, sinkers, floats and hooks, fish weirs and traps. The recognition of such evidence is consequential upon two factors which are fundamentally taphonomic in nature: the probability of preservation and the rate of discovery.

The recovery of fishing artefacts in the archaeological record in Scotland, however, is quite poor. In a land with a high degree of humidity and predominantly acid soils, wood, ropes, nets and other organic materials will decompose rapidly. Survival of inorganic material in recognisable forms is much more frequent, although this is often qualified by a change in use or re-use at a later period. If fishing was a major activity, stones may have been used as sinkers and metal as hooks. There is no evidence for iron hooks until the Viking period, but this may be due to the general scarcity or re-use of metal until that time.

Therefore, the primary evidence for fishing in Scotland during prehistory is the presence of fish remains and the abundance of such assemblages (see Barrett *et al.* 1999). In our endeavour to interpret and assess the relative importance of such remains, we have had to consider

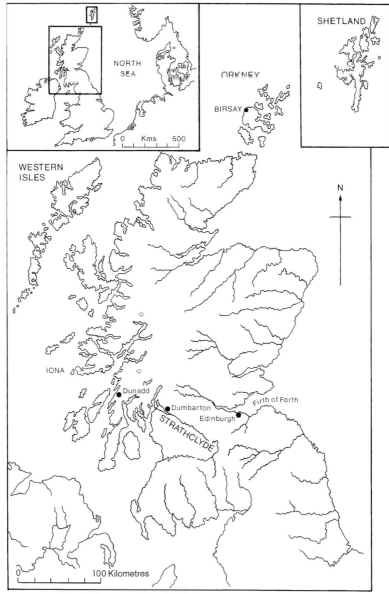

Fig 8.1 Map of Scotland showing the different regions.

elements of the complex ethnographic/historical record that are potentially available to us in Scotland as well as the obvious modern biological studies.

Primary information was also sought by interviewing retired fishermen, particularly from the Western Isles. These fishermen sailed in small open boats as young teenagers, like their fathers before them and prior to the introduction of steam vessels. It is worth noting that fishing communities around Scotland, until recently, were singularly close-knit; the sons of fishermen would almost invariably become fishermen themselves (Martin 1995), and knowledge of fishing grounds, species and fishing methods was passed on from generation to generation.

Geographical Set Up

The coast of Scotland (Fig. 8.1) is deeply indented by inlets from the sea. The larger inlets are called firths. Long, narrow inlets in the west are called sea lochs. On the rugged west coast these sea lochs are framed by great cliffs and at times resemble the fjords of Norway. Numerous islands line the coast. In the north are two large groups, the Orkney Islands and the Shetland Islands. Close to the west coast are the Inner and Outer Hebrides groups, and the islands of Arran and Bute.

The Outer Hebrides (also called the Western Isles) consist of a series of islands which lie off the north-western coast of mainland Scotland. They extend over a linear distance of some 200km and have a combined coastline over 1,800km long. The most northerly is the island of Lewis and Harris. This is some 65km long and approximately 20km to 40km wide. It is the largest and presently the most heavily populated island in the Outer Hebrides, although rural areas continue to be scantily populated. One of its main features is the 'Black Moor,' *i.e.* the boggy peat lands which blanket most of the interior of Lewis.

Traditional Fishing Methods

Fishing From the Shore

Rock fishing was an important method of fishing, particularly in the islands, and was usually carried out with a frame net, known in the Western Isles as *tabh*, i.e. 'bag-net' (Fig. 8.2). In the Western Isles one of the materials used for the netting of the *tabh* was the carnation grass root (*Carex flacca*). The fish caught with the *tabh* were mainly young saithe (*Pollachius virens*), which were known by different names according to size and age (in the Western Isles saithe were known as *cudaigh* or *cuddy* in their first year, *smallach* in their second year, and *saoidhean* in their third year).

The number of names attached to saithe throughout Scotland, in its various stages of development are indicative of its undeniable value in the diet of its population since the Mesolithic.

The same type of frame net was employed in Orkney. There it was called a *sillock-pocks* (*sillock* being the name given to young saithe in the Northern Isles) and was made in the form of an umbrella suspended at the end of a long pole. Such nets ranged from 2.5–3m in diameter and would be used in water less than two fathoms (1.8–3.6m) deep. Bait, mainly of pounded limpets, would be thrown over the net to attract young saithe (Fenton 1978).

Rock fishing with rods was also practised, sometimes alongside the frame net (Fig. 8.3). The rock fishing rod (*slap creagach*) was used mainly for catching first to fourth year saithe. Other species caught by using these methods included mackerel (*Scomber scombrus*), wrasse (Labridae) and young cod (*Gadus morhua*). Minced bait would be thrown off the rocks to attract the fish. The bait consisted mainly of lugworm, herring, mackerel, crab, whelks, mussel and other crushed shellfish.

Leepits

Crushed limpets, referred to as *soll* in the Western Isles, were used specifically in fishing from rocky areas. Limpet holes or *leepits* (Fig. 8.4) would be carved into rocks to keep a supply of *soll* during rock fishing. Spence (1899) describes these as "cup-holes" in Shetland, where they

Fig. 8.3 Working a 'poke-net' for saithe from the rocks. Shetland. Although it is unlikely that the boy with the rod would have been casting so close to the net fisher (Martin 1995, 9).

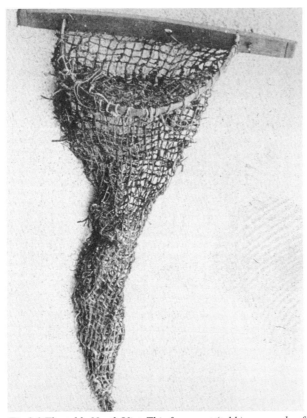

Fig 8.2 The tabh. North Uist. This frame net (tabh) was made of grassroots (Carex flocca). It was 130cm long; the opening was approximately 50cm (Photo by E. Beveridge 1911; Copyright School of Scottish Studies, University of Edinburgh. BIII b2 3.3 6619).

Fig. 8.4 Limpet pits. Lochboisdale, South Uist. These were artificial hollows, cup-holes and basins in rocks which were used for crushing limpets and other shellfish used for baits (Copyright School of Scottish Studies, University of Edinburgh. BIII 3 f 1652).

were connected to *craig* or rock fishing for saithe. The purpose of these was for holding *rooder for soe*, i.e. lure or bait. The bait used for catching young saithe off rocks in Shetland, as in Lewis, was mainly chewed or crushed limpet. Spence also points out the possibility that the cup-holes were also intended as a sign of proprietary right to particular *craig* sittings that may also have been passed on from generation to generation.

Martin (2004) describes the leepits as "rock-hollows" that were used on the island of Gigha in Argyll and Bute for rock fishing. In such rock-hollows, limpets would be crushed and used as bait for fishing lines and/or as ground bait that would be thrown on to the sea to draw in fish. These rock-hollows were approximately 20cm across and 10cm deep (Martin 2004).

Fishing at Sea

Boats

Maritime technology has been vital to the basic subsistence of the Scottish islands, where the surrounding seas have been fecund in contrast to the agriculture potential of the land, which is limited by rigorous climatic and geological factors. Boats were developed when there was a need for them and then became essential for people to survive. It is speculated that the first boat types in the Hebrides may have been equivalent of sea-going-boat shaped *currachs*, or coracles, made of wickerwork and animal hide. The earliest written reference to coracles in Scotland derives from Adamnan's *Life of Columba*, Book II, after 679 AD (Anderson 1990; 1991) and from the *Historia Scotorum* (*A Latin history of Scotland*) by Hector Boece (1527), where he describes a boat made from wands and bull hide called a *corrok* (currach), which was used for fishing salmon and was easily carried on people's backs.

People did not take to the sea unless the benefits for subsistence were considerable. If the land alone could not support the human population, men also took to the sea for food. Thus, boats developed and time, work and scarce raw materials were put into making them. Boats evolved in different ways throughout the world; their development was conditioned by the geography of local waters, climate, and purposes for which the boats were needed, as well as the availability of raw materials (Wilson 1965; Jarman *et al.* 1982; McKee 1983; Oslen 1983; Morrison 1992; Greenhill and Morrison 1995; Tanner 1996; Rixson 1998). Against this background, the Scottish islands have a long tradition of seafaring in small open boats, using sail and oar power to get to and from the fishing grounds, and for transporting people, animals and goods around the islands; these boats were a part of everyday life (Fig. 8.5).

Before the commercialisation of fisheries, many people living on the coast were crofters as well as subsistence fishermen, catching only enough fish for their own needs and using small open boats. White fish (*e.g.* saithe and cod) and herring taken by line were caught only in sufficient quantities to satisfy local demand (Wilson 1965). In Lewis the *yole* or *geola* was used for inshore fishing while the

Fig. 8.5 Arisaig, South Uist, 1953. Boat carrying people from island to island, probably for a Sunday Service or Mass, judging by the formal dressing (Photo by J. Lennie 1953. Copyright School of Scottish Studies, University of Edinburgh C12 b4 2042).

Fig 8.6 Yole from Fair Isle, 1920 (Photo by A. Fenton 1962; Copyright School of Scottish Studies, University of Edinburgh. BIII d3 1 7152).

Fig. 8.7 Sgoth boat type. St Kilda, Western Isles (Photo by G. W. Wilson before 1889; Copyright School of Scottish Studies, University of Edinburgh. C 139 b4 1838).

sgoth (skiff) was a larger, sturdier vessel used for deep water fishing for cod and ling (*Molva molva*). *Yoles* or *geola* were small open boats of under 20 feet (6m) keel. These were used for the small-line fishing in Lewis and would usually carry a crew of two men (Fig. 8.6). The *sgoth* (Fig. 8.7) was the Hebridean open, sail-powered, wooden boat

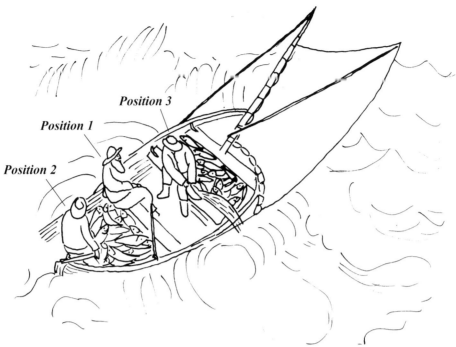

Fig. 8.8 Fishing arrangements in the small open boats (Drawn by the author based on a painting by the Icelandic artist Finnur Jonsson).

of up to 35 feet (11m). It originated from North Lewis and could take a crew of six to seven. The *sgoth* was famous for its ability to sail the hostile waters around the islands. It had to be light enough to be operated from open beaches but strong enough to withstand the rough Hebridean Sea (Tanner 1996).

Each *sgoth* carried a fire kettle which consisted of a three-legged iron pot with burning peat inside. The *sgoth* fishermen used great lines to catch cod and ling over the winter months from fishing banks extending from northeast of the Butt of Lewis to opposite Stoer Point on the east side of the Minch (Miller 1999). The *sgoth* would go to sea from Monday mornings to Friday evenings, and they were mainly used for catching cod, particularly in late spring (when young cod started to leave the fjords for deeper waters) and also for fishing during autumn. Hand lines, long lines and nets would be used.

This ethnographic description shows how the crew's tasks were organised (Fig. 8.8). The outline of the traditional fishing boat shows how the different tasks were carried out and where the orders derived from. The task of the man in the second position was to hold the boat in place with the oars. His instructions came from the skipper who sat at the stern, in position one. The skipper put the line over at the same time as he took the fish and threw it to the third man, position three, who assisted by distributing the catch into the holds.

All boats, even those that work well offshore, depend on having a refuge that can be reached in bad weather. This may be a harbour or even a beach if the boat is the right shape. Shingle beaches were favoured by fishermen because this facilitated the task of beaching the boat (McKee 1983).

Any culture is a mixture of old and new, and traditions endure when they can function under new conditions. In the fishing community 'new' elements consist in the introduction of motorised and larger boats, particularly after the First World War.

A fisherman must know how to handle his boat and his gear and where to allocate the fish. Fishing from boats at sea has involved several methods.

Line Fishing
Before the 1880s, long-lining was the usual method used to catch white (demersal) fish such as cod, saithe, ling and flatfish such as the halibut (*Hippoglossus hippoglossus*) which live at the bottom of the sea. It was very labour intensive but resulted in a high quality catch. There were two types of long-line fishing: the small-line method (*sma line*) and the big or great-line. The small-line was carried out using the small boats *i.e. yeolas*. Lines consisted of lighter strings (in weight) than those used for the great-line fishing (see below).The small-line generally consisted of horse hair snoods (also known as *snuids* and *tippins*) and was 150 fathoms (*c*. 275m) long with up to 250 baited hooks suspended from it by the *snuids*. Bait used for the small-line were shellfish: mussel (*Mytilus edule*), limpet (*Patella vulgata*), lugworms, sandeel (*Ammodytes*), herring (*Clupea harengus)*, mackerel and crab. The fishes caught with the small-line were haddock (*Melanogrammus aeglefinnus*), whiting (*Merlagius merlangus)*, gurnard (*Eutrigla gurnardus)*, young cod, thornback ray (*Raja clavata)*, plaice (*Pleurenectes platessa)*, flounder (*Platichthys flesus)*, sole (*Solea solea)* and dogfish (*Scyliorhimus caniculus)*.

Small-line fishing was a winter and spring activity, frequently continued into the summer and often worked

Fig 8.9 Preparing horse hair for making lines and/or ropes. Lochboisdale, North Uist (Photo by W. Kissling 1936; Copyright School of Scottish Studies, University of Edinburgh. BVIII3 c 5449).

Fig 8.10 Twisting the horse hair to make fishing lines. Lochboisdale, South Uist (Photo by W. Kissling 1936; Copyright School of Scottish Studies, University of Edinburgh. BVIII3 c 5452).

Fig. 8.11 Illustration from the island of Eriskay (Western Isles) of rope making using heather (Calluna vulgaris) (Goodrick-Freer 1902; Copyright School of Scottish Studies, University of Edinburgh. B VIII 3 c2883).

consecutively with great-line fishing. At the beginning of the season, boats would remain close inshore within 1 to 5 miles, but as the season changed the boats went further offshore. By May they would be 8–9 miles, c. 15–16km, out to sea. Small-line fishing was a family affair with women and children responsible for preparing the equipment, and this was the case in most fishing communities throughout Scotland up till at least early 20th century.

Great-line fishing was similar to small-line fishing but was undertaken in deeper waters further out to sea, and was carried out using the larger boats *i.e.* the *sgoths*. The long lines were baited and set for up to 48 hours, fixed with stone anchors. These were formed by seven strings of 60 fathoms each (110m), making it an average length of 420 fathoms (770m), with each line carrying approximately 168 hooks. Stones were used to sink the great-lines to the bottom. The bait used for great-line fishing was conger eel, haddock, herring, mackerel and occasionally squid. The fishermen usually baited the lines on the boat. Deep-sea fishes caught with great-lines were ling (*Molva molva*), cod, tusk (*Brosme brosme*), hake (*Merluccius merluccius*), turbot (*Scophthalmus maximus*), halibut, conger eel, dogfish, sea angler (*Lophius piscatorius*) and shark.

Because of the work involved in preparing and hauling the lines, new methods of catching white fish were sought. Trawling was introduced into Scotland from England in the late 19th Century. In the 1920s, seine netting was introduced from Denmark, thus replacing the traditional methods, *i.e.* the 'small' and 'great' lines.

Fishing Implements Employed in Boats

Fishing Lines and Nets

Lines made from hemp had snoods or *snuids* made of horse hair from which baited hooks would usually be attached at 6-foot (1.8m) intervals. A line would be hauled from 30–40 fathoms (55–73m) of water. Great stamina and strength were needed for such an arduous task. As a line was hauled, another fisherman would unhook the fish while separating them by species and size. Lines were cleaned during the afternoons; hooks would be replaced and lines cleared of other organisms such as starfish and seaweed. In the case of the small-line method, each fisherman would take his own line home for baiting.

In the islands, fishermen made fishing lines from twisted horse hair (Figs. 8.9 and 8.10). Ropes were made of heather and hay. The processes for making lines and ropes, mainly by twisting and plaiting, are similar throughout the Northern and Western Islands (Figs. 8.11 and 8.12).

Fig. 8.12 Rope making with heather (Calluna vulgaris) in Lochboisdale, South Uist (Photo by W. Kissling 1947; Copyright School of Scottish Studies, University of Edinburgh. B VIII 3 c 5444).

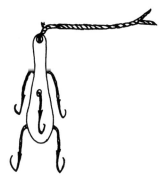

Fig. 8.13 Deep-sea dorgh (dorgh mòr). (Drawn by the author from a drawing by G. MacLeod displayed at the museum in the Comunn Eachdraidh Sgire Bhearnaraidh, Bernera Historical Society. Not to scale.)

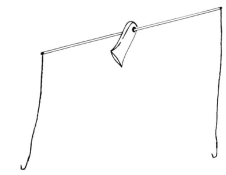

Fig. 8.14 Hand-line dorgh (dorgh-beag). (Drawn by the author from a display in the museum of the Comunn Eachdraidh Sgire Bhearnaraidh, Bernera Historical Society. Not to scale.)

The horse hair used was from stallions and geldings, never from mares, though the preferred hair for the manufacture of fishing lines was that of ponies. According to ethnographic accounts from the Western and Northern Islands, black and grey hair made better fishing lines. The hair was washed and left to dry at a certain temperature; as depending on its moisture, hair would make the lines too tight, or too springy. The types of fish caught with horse hair line used for fishing from rocks or boats included sea bream (*Pagellus* sp.), whiting, cod and saithe (Henderson 1968). Before the mid-nineteenth century, nets were made from hemp (*Cannabis sativa*). The use of hemp for fibre has been traced to the Bronze Age in Scotland (Ryder 1999). Fishermen also produced their own netting needles made from wood or animal bone (Martin 1981).

Deep-sea fishing with ripper (*dorgh mòr* or *dorgh*) (Fig. 8.13) was another method used from boats. This method of fishing was referred to as *dorobhaigh* in the Western Isles. The *dorgh* was made by casting approximately 7lbs (3.5kg) of lead into a sand mould or into a circular hole in the ground. It was then polished and one end flattened. Three holes were then drilled for the *snuid* (*i.e.* the snood), and one hole was made in the flat end. Prior to use, the *dorgh* would be scraped with a knife to make it shiny (Macleod undated). The *dorgh* would be let down from the boat until it touched the bottom of the sea. It would then be hauled back approximately 1 fathom (1.9m) and worked up and down at that depth. No bait was used since the shining *dorgh* would attract the fish, usually cod, which would jump at it and thus get hooked. Other fish caught with the deep-sea ripper *dorgh* were saithe and pollack, turbot and halibut (Mcleod undated).

A hand-line *dorgh* (*dorgh-beag*) (Fig. 8.14) was used to fish from boats inshore. This hand-line *dorgh* consisted of a weight of approximately 4lbs (2kg), a wire with 4 to 6 *snuids* or snoods with hooks, and a hand line of approximately 60 fathoms length (110m). The fish caught with the hand-line *dorgh* were the same as those caught with the small-line. This fishing tackle was used inshore and the hooks were baited. The *dorgh* would be let down and the moment of catch would be felt on the hand line which would then be used to haul the fish aboard (Mcleod n.d.).

Buoys Used in Great-Line Fishing

The great-line buoys were made of sheepskin in the Western Isles (Fig. 8.15), and similar buoys were used in other areas of Scotland until the beginning of the 20th century. The sheepskin buoys were made by soaking the skin in water until the entire fleece had been shed. The maker would then gather the edges of the skin, fold them about the base of a wooden stock, and lash them with marline, a light, tarred rope, made of two strands. The skin would then be inflated and judged by the roundness until the intended shape was achieved. The skin would be covered repeatedly with marline, tightening it firmly. When thoroughly dried, a mixture of tar and linseed oil would be spread inside the buoy, soaking the interior and thus making the buoy waterproof. The skin was kept soft and pliable so that it would not crack during warm days. On the east coast, dog skin and pig bladder were also used for making buoys, and cattle bladders were also used in some areas (Martin 1981; Wigan 1998). Buoys made from animal skins were used

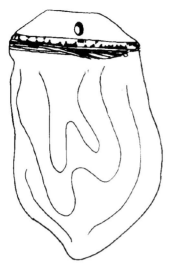

Fig. 8.15 Sheepskin buoy (Pùb-craiceann). (Drawn by the author from a display in the museum of the Comunn Eachdraidh Sgire Bhearnaraidh, Bernera Historical Society. Not to scale.)

until the 1920's in some areas, *e.g.* Eriskay in the Western Isles, until canvas buoys were introduced (Fig. 8.16)

Use of Landmarks and Fishing-Marks

Landmarks were an essential source of direction for fishermen before the use of the compass, although their use still continued even after its introduction. Such landmarks allowed the fishermen to 'draw' fishing-marks which were used to find particular places that certain fish populations were known to frequent and, similarly, to avoid areas where good fishing was unlikely (Eunson 1961). Some of the fishing grounds may have been close inshore, the majority situated between 1.5 and 5km from the shore, and the farthest distance for fishing was no more than 10km. Time and tide were the most important elements in fishing. Time and weather were never far from the thoughts of fishermen in the past (Eunson 1961). Landmarks were physical features on the land or in the sea: headlands, prominent rocks, islands, and even houses. These 'marks' also enable the fishermen to know the position of the boats and relate it to the different fishing grounds or fishing banks.

In Lewis the landmarks became so standard through the years that they governed the areas fished from different fishing villages in the island, *i.e.* the landmarks defined the fishing grounds exploited by different settlements in the island.

In Great Bernera, Lewis, for example, the inshore winter and spring mark was Sgeir Rhudha which is an area of rock and coral stretching north westerly out from Gallow Head for approximately 5km. This was a good winter and spring fishing ground for ling, cod and skate, particularly in the area of Bealach Skipedail ('the pass of the ship-valley'), a low lying hollow between Ard Mòr Magersta and Eilean Molach. Fishermen became familiar with the type of seabed from the types of fish that came up on the hooks. Shallow

Fig. 8.16 Canvas buoys (boughs) replaced those made from animal skins. The wooden plug tied tightly at the top in the same manner as those used in buoys made of animal bladders or skins. Fife 1900 (Photo by fisherman Mr. John Reid [McGowan 2003]).

spots were discovered when the line was suddenly caught on the bottom. When the lines came up badly twisted, it usually meant that they were caught up in a swift flowing eddy or current. Treacherous seabed, which could mean the loss of fishing lines, was often marked by swift dangerous currents on the surface (Eunson 1961).

The great-line technique could not function in the type of hard ground such as that of Sgeir Rhudha (after MacLeod n.d.) therefore small-lines must have been used.

The boats would run out to sea from the point of an 'out-going mark' until they reached a 'cross' or 'tangent side mark.' In this way fishermen could trace the marks if the lines broke.

One outgoing mark used by Bernera fishermen was Conostom Hill, and as boats moved outwards to the marks more headlines were needed. For example, for great-lines, the Great Bernera fishermen took their marks off the Flannan Islands (Fig. 8.17). Macleod (n.d.) gives two reasons for this, the fishing grounds of the Great Bernera fishermen were always located to the East and North East of the Flannans and secondly, these islands would be in sight even if the inner land mark was totally obscured. Six marks were used for the outer fishing grounds (see Fig. 8.17):

1. Carlais Lamh a Sgeir (Sgeir is an inlet between Eilen Mhor and Eilean Tigh);
2. Tir a Muigh (the three outer isles of the Flannans);

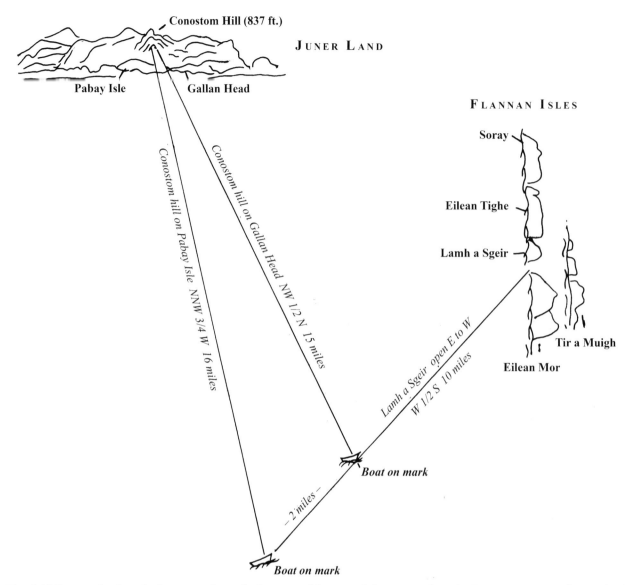

Fig. 8.17 Bernera land marks from great lines: furthermost of the outer fishing ground marks. Perspective view of Juner land and Flannan Isles as seen from the marks. Conostom Hill: kept on Pabay Isle or Gallan Head. Lamh a Sgeir and Eilean Mor separated by a channel (Drawings by the author based on those by G. MacLeod).

3. Starr (a rocky conical islet to the west of Eilean Mhor);
4. Rhu Breach Soray (Northwest point of Soray island);
5. Carna Eilean Thigh (a stone stock in *Eilean Tigh*), and;
6. Soray island.

Division of Labour in Fishing Communities

Fishing has commonly been considered as a 'man's work'. In most fishing communities the division of labour by gender and age indeed tends to be quite sharp. The masculine image of the industry, however, conceals the reality of this occupation: removing men to sea, makes them particularly dependent on the work of women, old men and children ashore (Figs. 8.18 and 8.19). The conventional image of fishing as a male occupation has thus grossly misrepresented and undervalued the real part played by women in the industry. Women almost everywhere have made a central contribution to fishing through providing logistical support on land, both before and after fishing. Work ashore in preparation for fishing and disposing of the catch was left entirely to the women or shared with them. Work at sea, however, was reserved for men. This feeling is so strong that in many places women would not be allowed on a boat about to set off to sea. Furthermore, in many places the presence of women on vessels was considered a bad omen (Anson 1950; Thomson *et al.* 1983; Thomson 1985).

In most communities the men caught the fish and they normally relied on women both in preparations on shore and, more generally, in disposing of the fish caught. For example, the inshore line fishery relied on women for the gathering of bait, and cleaning both bait and lines. After the catch was landed, women would either split the fish for

Fig. 8.18 Children making nets, South Uist (Photo by Shaw 1935; Copyright School of Scottish Studies, University of Edinburgh. B III 3 c2 4318).

Fig. 8.19 Dividing the catch on shore. 1880s Auchmithie, Angus, East Coast. In small-line fishing a share went to each of the crew with an extra share sold for the upkeep of the boat (King 1991).

drying or sell it fresh. This type of household production remained dominant until line fishery was replaced by trawling and seining between the 1880s and mid-20th century (Thomson 1985). The type of line fishing described here was in decline by the end of the 19th century owing to the impact of steam trawling, although ground seining did not appear on the west coast until after World War II (J. Coull pers. comm. 2002).

The Interpretation of Archaeological Fish Bone Assemblages in Scotland

Jochim (1976) pointed out the importance of choices in a prehistoric society. This includes the choice of which resources to utilise, and decisions as to their proportional use and the time accorded to them. These factors require time and energy and can be envisaged to structure the subsistence pattern of human settlements. Another factor concerns the nature of choices: whether they are the result of long-term planning and the consideration of goals or are more opportunistic. Another important point in the exploitation of natural resources is risk minimisation. Environmental factors affect performance, therefore climate, geography and seasonal conditions must be taken into account in the approach to risk. Settlement location is also important, and it has long been recognised that the form, size and permanence of human settlements bear a definite relationship to the modes of exploiting the natural environment (Murdoch 1969; Jochim 1976; 1981).

Example from the Western Isles: Bostadh in Great Bernera, Lewis

Bostadh is situated in Great Bernera, a small island off the West Coast of Lewis. This island is 7km long and 3km wide and is now reached by a road bridge across the Atlantic Ocean from the Isle of Lewis. Prior to the construction of the bridge in 1952, this island was reached by boat. Bostadh is located within what is a presently rapidly eroding *machair* (Figs. 8.20 and 8.21). The shell component of *machair* deposits allowed an extraordinary degree of preservation of archaeological environmental material,

Fig. 8.20 Bostadh in Great Bernera, Lewis in a summer sunny day.

Fig. 8.21 Bostadh in Great Bernera, Lewis in a winter, stormy day.

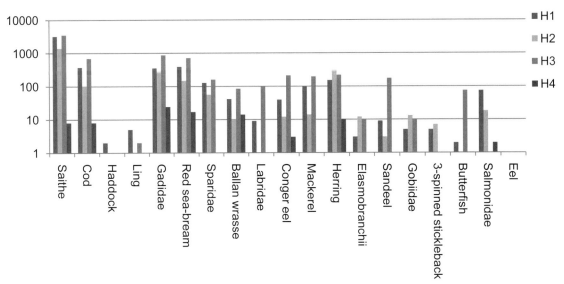

Fig. 8.22 Bostadh Later Iron Age fish species representation by house (using NISP). H = house.

which, coupled with the surviving upstanding architecture within the now collapsing shell-sand component, has provided a rich archaeological source of well preserved environmental assemblages, of which fish and shellfish remains were most abundant.

Bostadh represents the remains of a settlement sequence from the Late Iron Age, c. 400 BC to c. AD 800, to putative early Norse, c. AD 800 to c. AD 1050 (Neighbour 2001). While this sequence is common in Orkney, very few Hebridean sites with such chronology have so far been recognised, the exceptions being represented by Udal and Bornish in North Uist (Crawford 1996; Sharples 2005). This may be mainly due to the loss of coast through erosion.

Late Iron Age

The analysis of the fish remains recovered at Houses 1, 2 and 3 of the Late Iron Age settlement at Bostadh revealed that the main species represented in the assemblage is immature (first and second year) saithe (*Pollachius virens*). Other species present included cod (*Gadus morhua*), red seabream (*Pagellus bogaraveo*), mackerel (*Scomber scombrus*), conger eel (*Conger conger*), ballan wrasse (*Labrus bergylta*), herring (*Clupea harengus*) and Salmonidae (salmon *Salmo salar*/trout *Salmo trutta*) (Fig. 8.22).

The large amounts of fish remains recovered from sediments of the Late Iron Age settlement give evidence of specialised fishing practices during this period on the island of Great Bernera. Gadids were the main group exploited with a predominance of immature saithe over cod. Gadidae, fishes of the cod family, have been of particular importance throughout pre- and proto-historical periods in northern Scotland. Cod and saithe dominate almost all of the northern Scottish fish bone assemblages (Barrett *et al.* 1999). The dominance of saithe over cod in the Bostadh assemblage is characteristic of Late Iron Age sites in other areas of Scotland. Saithe are represented in most Iron Age sites throughout Scotland are generally smaller than cod.

Fig. 8.23 Blackhouse. Loch Eynort, South Uist 1936. The Hebridean blackhouse was a turf-walled thatched house (Copyright School of Scottish Studies, University of Edinburgh A VII 3b 8084).

The specimens usually represented are first and second year-old specimens, measuring between 15cm to 30cm in total length, while cod are generally more mature specimens measuring from 30cm up to 100cm in total length. In the Late Iron Age Bostadh assemblage, fishes of both these size groups are well represented.

Saithe are found in shallow water particularly during their first 3–4 years. During this time they range from 15cm to 55cm in total length, according to modern fisheries data. In the Scottish islands in the recent past these were caught by net (*tabh*) or simple rod and line from the shore and from boats in shallow water. Saithe can be eaten whole (once gutted), sometimes with the liver, but they can also be smoked and thus preserved for later use. This has been a practice in the Hebrides where small fish were simply hung inside the blackhouses (Fig. 8.23), where they would be smoked by the domestic peat fire.

It is also interesting to note that young saithe have traditionally been used in Scotland for the extraction of

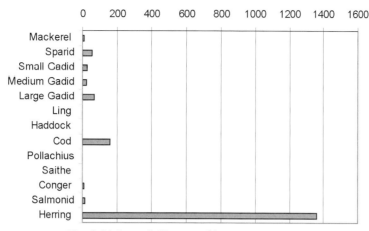

Fig. 8.24 Bostadh Norse midden representation.

fish liver oil (Smith 1984). It is recorded that in Skye for instance, the oil used for lamps or *crusies* was extracted from the liver of fish caught for domestic use (McGregor 1880, 145).

Cod has been traditionally caught by hand-line both from the shore and from small boats within sight of land. Cod elements recovered at Bostadh were mainly from specimens in the range of 30cm to 60cm in length, though some elements from larger specimens (60cm to 100cm) were recovered, mainly from House 1. The presence of medium (<60cm total length) and large (<100cm total length) cod that live further offshore indicates that boats were employed for their capture, using baited lines.

The non-Gadoid taxa represented in the assemblage are herring, red seabream, mackerel, conger eel, wrasses and Salmonids. These occur in other sites around Scotland, and despite their small numbers when compared to the Gadids, they are relatively consistent. The red sea bream was present in the Bostadh assemblage. This species belongs to the Sparidae family that, although recorded in assemblages ranging from the Neolithic to the Middle Ages in Scotland, has generally occurred in relatively small numbers in Scottish fish bone assemblages. The red seabream inhabits relatively deep inshore waters over a variety of seafloor environments (Whitehead *et al.* 1986) and would require line fishing.

Mackerel were also present in this assemblage. This common North Atlantic fish is found seasonally close inshore. Mackerel is an important food fish and is caught in various ways including nets and by hook (Wheeler 1969; 1978).

Herring is found on the west coast in winter moving towards Shetland in the spring and summer. During the winter months, the west coast islands and mainland benefit from a different population derived from the Atlantic sources (Gray 1978). Herring require netting for their capture although they can sometimes be caught by line.

The Norse Period
Excavations at Bostadh revealed a settlement of figure-of-eight houses, sealed by a Norse Age structure and associated midden. This evidence has been compared with that from other sites. It is argued that there is convincing evidence for a significant change in the economy in the Norse period, a change also revealed by the analysis of human skeletal material in other areas of the island of Lewis, such as Galson Beach (Neighbour *et al.* 1998).

The main species present in the Norse midden is herring (*Clupea harengus*). Vertebrae were the only elements from this species and were found to belong to mainly mature specimens of approximately 35cm total length. Herring can grow up to 45cm. Herring has been one of the most important species in Scotland's fishing history. The migratory habits of the herring are unusual and Scotland is particularly fortunate in relation to its movement. For example, Loch Roag in Lewis is open to the waves from the Atlantic and before the mid 18th century it was well known for its rich herring grounds (Mitchell 1864, 58).

Gadids were the second most represented group recovered in the midden; cod, saithe and ling were the species identified. Cod was the second largest species present in this context with mainly medium (30cm to 60cm total length category specimens) and large (60cm to 120cm total length category specimens) animals represented (Fig. 8.24). For a comparative resume of the fish species recovered at Bostadh and their relative sizes and mode of catching refer to Table 8.1.

Conclusion

The main purpose of this work has been to emphasise the importance of understanding one's environment, focusing on the direct interaction of the people of coastal Scotland with the sea. It has attempted to expose the close relationship the inhabitants of this region had with their environment, where even language itself developed according to observation by people of their surroundings, particularly Gaelic in the Western Isles. It was also written to enhance the recognition of one element in the anthropology of human exploitation of the environment in this part of the North Atlantic region, and to provide an example of a comparative, multidisciplinary study linking

Species	Size range (cm)	Habitat (juvenile)	Habitat (Adult)	Spawning Season	Possible Fishing Method for juvenile specimens	Possible Fishing Method for adult specimens
Cod	up to 150cm	Shallow water/ close inshore	Inshore and offshore/ depths down to 600m	February to April	Line from craig-seats, poke net (*tabh*)	Trailing line from craig seats. Lines from boats
Saithe	up to 100cm	Shallow water close inshore among rocks	Shallow and deep water	January to April	Poke nets (*tabh*), lines from craig seats	Lines from craig seats or boats
Pollack	up to 130cm	Shallow water close inshore among rocks	Shallow and deep water	January to April	Poke nets (*tabh*), lines from craig seats	Lines from craig seats or boats
Haddock	up to 100cm	Mainly offshore	Offshore depth from 10 to 450m	January to June	Lines from boats	Lines from boats
Ling	up to 100cm	Shallow water	Offshore/depths down to 400m	March to July	Lines from craig seats or boats inshore	Lines from boats offshore
Conger	up to 150cm	Inshore	Inshore	Summer in Sargaso & Mediterrenean Seas	Lines from boats	Lines from boats
Wrasse	up to 30cm	Argal zone on rocky coast	Argal zone on rocky coast	June to August	Lines from craig seats or boats inshore	Lines from craig seats or boats inshore
Red sea bream	up to 50cm	Coastal areas in shallow waters	Offshore banks in deeper waters down to 700m	November to February	Lines from boats	Lines from boats
Herring	up to 40cm	Pelagic, depths down to 250m	Pelagic, depths down to 250m	February to April in Norwegian coast	Nets from boats, inshore fish traps	Nets from boats, inshore fish traps
Mackerel	up tp 50cm	Pelagic, coastal waters	Pelagic, coastal waters	May to June	Lines from craig seats or boats	Lines from craig seats or boats
Salmon	up to 150cm	Freshwater then estuarine to acclimatise to salinity in sea	Open sea and freshwaters	Spawning migrations from sea into freshwater June to November	Lines from boats	Lines from boats, nets and fish traps

Table 8.1 Notes on species habitats and possible method of capture.

archaeological evidence with ethnography and the natural sciences.

Thus, fishing methods as well as the different implements used for catching different species have been described, including the types of man-powered boats and the use of natural landmarks for directing the crews to and from the shore to the fishing grounds. This account stresses the close interaction between people and their environment, before technology such as compass and motor-powered boats helped fishermen to either venture out to sea safely or return to the safety of the land. It has also been stressed that activities such as fishing require organisation and knowledge of the natural environment which is being exploited. Fishermen must know where and when to find the fish (*e.g.* what time of the year) and how to catch it. It is an activity which relied on the work and support of the whole community.

Bostadh is just one example of the many sites which have been analysed bearing in mind the ethnographic background of fishing in Scotland. As in Bostadh, most of the fish bone assemblages analysed in this region so far have been interpreted as exploitation of the immediate environment around the sites from which fish remains have been recovered. In the case of Bostadh, the rocky surroundings would have allowed for an extensive exploitation of immature saithe. The sandy beach would have allowed safer access to offshore fishing with hand lines for large gadids and other offshore species identified in the assemblage, whilst herring would have been an easily accessible resource due to its migratory habits that formerly benefited these islands. Ethnographic accounts have therefore played a major part in our understanding of fish remains recovered in archaeological sites in Scotland, making a very significant contribution to our reconstruction of ancient fishing and economic practices.

Acknowledgments

This author would like to thank The School of Scottish Studies, University of Edinburgh, for their valuable assistance, in particular Mr. Ian MacKenzie for providing images from the Photographic Archive. Special thanks are also given to the Bernera Local History Society (*Comunn Eachdraidh SgireBhearnaraidh*) for providing manuscripts by the late George Macleod of Breaclete whose handwritten accounts on the fishing lore, methods and past fishing practices of the island of Bernera are 'a heritage and gift of knowledge of past generations.' Thanks also to Professor Ian Ralston, Dr. Noel Fojut and Roderick McCullagh who read and commented on earlier versions of this work. Thanks are also given to the British Academy for their financial support which made possible assistance to the 10th ICAZ International Conference where this paper was first delivered.

References

Anderson, A. O. (1990) *Early Sources of Scottish History AD 500–1286*. Stampford, Paul Watkins.

Anderson, A. O. and Anderson, M. O. (1991) *Adomnán's Life of Columba*. Oxford, Cavendon Press.

Anson, P. (1950) *Scots Fisherfolk*. Banff, Banffshire Journal Ltd. for the Saltire Society.

Barrett, J. H., Nicholson R. A. and Cerón-Carrasco, R. (1999) Archaeo-icthyological Evidence for Long-Term Economic Trends in Northern Scotland: 3500 BC to 1500 AD. *Journal of Archaeological Science* 26(4), 353–388.

Boece, H. (1527) *Scotorum historiae a prima gentis origine*. Paris.

Connock, K. D., Finlayson, B. and Mills, C. M. (1992) Excavation of a shell midden site at Carding Mill Bay near Oban, Scotland. *Glasgow Archaeological Journal* 17, 25–38.

Coull, J. R. (1996) *The Sea Fisheries of Scotland: A Historical Geography*. Edinburgh, John Donald.

Crawford, I. (1996) The Udal. *Current Archaeology* 147, 20–288.

Eunson, J. (1961) The Fair Isle Fishing Marks. *Scottish Studies* 5, 181–198.

Fenton, A. (1978) *The Northern Isles: Orkney and Shetland*. Edinburgh, John Donald.

Finlay, N. (forthcoming) Results of the Colonsay, Port Lobh excavations. University of Glasgow.

Goodlad, C. A. (1971) *Shetland Fishing Saga*. Shetland: The Shetland Times Ltd.

Gray, M. (1978) *The Fishing Industry of Scotland 1790–1914, A study of Regional Variation*. Studies Series 15. Aberdeen, University of Aberdeen Press.

Greenhill, B. and Morrison, J. (1995) *The Archaeology of Boats and Ships*. London, Conway Maritime Press.

Hardy, K. and Wickham-Jones, C. R. (2002) Scotland's First Settlers: The Mesolithic seascape of the Inner Sound, Skye and its contribution to the early prehistory of Scotland. *Antiquity* 2002, 825–833.

Hardy, K. and Wickham-Jones, C. R. (2009) Mesolithic and later sites around the Inner Sound, Scotland: The work of the Scotland's First Settlers project 1998–2004. *Scottish Archaeological Internet Reports*, 31.

Jarman, M. R., Bailey, G. N. and Jarman, H. N. (1982) *Early European Agriculture: Its Foundations and Development*. Cambrige, Cambridge University Press.

Jochim, M. A. (1976) *Hunter-Gatherer Subsistence and Settlement: A Predictive Model*. New York, Academic Press.

Jochim, M. A. (1981) *Strategies for Survival: Cultural Behaviour in an Ecological Context*. New York, Academic Press.

King, M. H. (1991) *An Auchmithie Album*. Arbroath, Herald Press.

Kissling, W. (1943) The Character and Purpose of the Hebridean Blackhouse. *The Journal of the Royal Anthropological Institute of Great Britain & Ireland* 73(½), 75–100.

Macleod, G. (n.d.) *Accounts of fishing lore, methods and practices in Lewis*. Handwritten Manuscript donated by Mrs H. A. Macleod to Lewis Castle College, Stornoway in 1969.

Martin, A. (1981) *The Ring-Net Fishermen*. Edinburgh, John Donald.

Martin, A. (1995) *Scotland's Past in Action: Fishing and Whaling*. Edinburgh, NMS.

Martin, A. (2004) *Fish and Fisherfolk of Kintyre, Lochfyneside, Gigha & Arran*. Glasgow, Bell & Bain.

McGowan, L. (2003) *Fife's Fishing Industry*. Stroud, Tempus Publishing Ltd.

McGregor, A. (1880) Notes on some Old Customs in the Island of Skye. *Proceedings of the Society of Antiquaries of Scotland* 14, 143–148.

McKee, E. (1983) *Working Boats of Britain*. London, Conway Maritime.

Mellars, P. (1978) Excavation an economic analysis of Mesolithic shell middens on the island of Oronsay. In P. Mellars (ed.) *The Early Postglacial Settlement of Northern Europe: An Ecological Perspective*. London, Duckworth.

Mellars, P. and Wilkinson, M. R. (1980) Fish otoliths as indicators of seasonality in prehistoric shell middens: the evidence from Oronsay (Inner Hebrides). *Proceedings of the Prehistoric Society* 46, 19–44.

Miller, J. (1999) *Salt in the Blood: Scotland's Fishing Communities Past and Present*. Edinburgh, Canongate.

Mitchell, J. M. (1864) *The Herring: Its Natural History and National Importance*. Edinburgh, Edmonton & Douglas.

Morrison, I. (1992) Traditionalism and Innovation in the Maritime Technology of Shetland and other North Atlantic Communities. In T. C. Smout (ed.) *Scotland and the Sea*. Edinburgh, John Donald.

Murdoch, G. P. (1969) Correlation of Exploitive and Settlement Patterns. *Bulletin* 230. National Museums of Canada.

Neighbour, T. (2001) Excavation at Bostadh Beach, Great Berneraigh, Isle of Lewis: Data Structure Report. Unpublished CFA Report.

Neighbour, T., Knott, C. and Bruce, M. F. (1998) Excavation of two long cists at Galson, Isle of Lewis, 1993–1996. Unpublished Report.

Olsen, A. G. (1983) *The Shetland Boat: South Mainland and Fair Isle*. Maritime Monographs and Reports No. 58. London, National Maritime Museum.

Rixson, D. (1998) *The West Highland Galley*. Edinburgh, Berlinn.

Ryder, M. L. (1999) Probable fibres from Hemp (*Cannabis sativa*) in Bronze Age Scotland. *Environmental Archaeology* 4, 93–98.

Sharples, N. (2005) *A Norse farmstead in the Outer Hebrides: Excavations at Mound 3, Bornais, South Uist*. Oxford, Oxbow Books.

Smith, D. H. (1984) *The Shetland Life and Trade 1550–1914*. Edinburgh, John Donald.

Spence, J. (1899) *Shetland Folk-lore*. Lerwick, Johnson & Greig.

Tanner, M. (1996) *Scottish Fishing Boats*. Shire Album 326. Shire Publications Ltd.

Thomson, P. (1985) Women in the Fishing: The Roots of Power between the Sexes. *Comparative Studies in Society and History* 27, 3–32.

Thomson, P., Wailey, T. and Lummis, T. (1983) *Living the Fishing*. London, Routledge & Kenan Paul.

Wheeler, A. (1969) *The Fishes of the British Isles and North-West Europe*. London, Macmillan.

Wheeler, A. (1978) *Key to the Fishes of Northern Europe*. London, Frederick Warne.

Whitehead, P. J. P., Bauchot, M. L., Hureau, J. C., Nielsen, J. and Tortonese, E. (1986) *Fishes of the North Eastern Atlantic and Mediterranean*. Paris, United Nations Educational Scientific and Cultural Organization.

Wigan, M. (1998) *The Last of the Hunter-Gatheres: Fisheries Crises at Sea*. Shrewsbury, Swan Hill Press.

Wilson, G. (1965) *Scottish Fishing Craft*. London, Fishing News.

9. Contemporary subsistence and foodways in the Lau Islands of Fiji: An ethnoarchaeological study of non-optimal foraging and irrational economics

Sharyn Jones

Drawing from my ethnoarchaeological work in the Lau Islands of Fiji, I discuss how modern studies of human–animal relations may contribute to archaeological questions. Archaeological research on prehistoric subsistence frequently utilizes optimal foraging (OF) models from behavioural ecology to explain human behaviour and to account for the zooarchaeological record. Data on subsistence and foodways from Lau are used to explore the usefulness of OF models for understanding human behaviour in both the past and the present. The underlying assumption of OF is that individuals relate to their environment in ways that maximize their reproductive success. Many common Lauan resource collection and production practices are non-optimal. That is, the behaviour goes against predictable foraging strategies with the best return for a given amount of effort. I contend that these behaviours are better understood in the context of local Lauan definitions of success and value. I identify and discuss four factors that affect or determine decision-making and planning prior to and during inshore fishing expeditions. These issues include: natural factors, the technologies accessible, the number of people in a fishing party, and the types of fish targeted. I also describe household-based consumption of fish and explain the material or faunal results and the archaeological implications of these practices.

Keywords: ethnoarchaeology, fishing, marine exploitation, optimal foraging, Fiji

Introduction

Broad and copious arrays of marine fauna make up a typical Pacific Island zooarchaeological assemblage. Understanding this material from an archaeological perspective alone is a difficult task. In this paper I draw from my ethnoarchaeological research in Fiji's Lau Group to illustrate how a modern study of human-animal relations and foodways may contribute to understanding of archaeological issues. Using these modern data I explore issues including the factors that influence people's decisions about fishing and eating. If the consumed fishes are a reflection of food preference or local availability, what portion of the diet does faunal material represent in terms of meals and the number of people fed, and so forth. In order to address these questions I combine detailed archaeological data with ethnographic observations focused on marine economic systems. With its contemporary population engaged in intensive fishing, the Lau Group offers an excellent opportunity to investigate long-term trends in marine resource exploitation.

The importance of inshore fishing to the economic systems of the Pacific Islands in both the past and present is undeniable. Indeed, it is almost impossible to understand economics and foodways on tropical oceanic islands without exploring inshore fishing practices. In Fiji's Lau Group, fisherwomen, adolescents, and men exploit the inshore area on a daily basis, collecting fish and shellfish that form the primary animal protein portion of their diet. Lauans plan their fishing trips and make decisions about marine exploitation based on a number of features including natural factors, the technologies accessible, the number of people in the fishing party, and the fish targeted. Together these elements structure behaviours and affect eating patterns. This research is an ethnoarchaeological exploration that seeks to account for why practices occur in the present and to hypothesize about human behaviour and societies that produced prehistoric archaeological material culture. Moreover, this work illuminates environmental influences on behaviours as well as the effects of social hierarchies on consumption patterns.

I argue that a historically informed archaeology sensitive to context can contribute much to the scientific

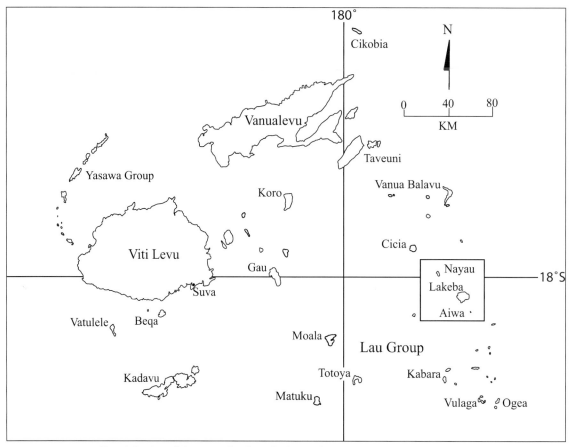

Fig. 9.1 Map of the Fiji Islands with the study areas indicated in the square

understanding of human behaviour. Ethnoarchaeology, as I practice it, is fundamentally comparative. I observe the social contexts in which material culture is collected, created, used, eaten, and discarded, with particular attention to physical fingerprints left behind by the sociological processes. This serves as a model, uniting social and material dynamics, which may then be used to search for similarities and differences in the archaeological record (see Schmidt 1997). The comparative analogical process leads to strongly supported inferences grounded in actual behaviour. Assumptions about how people should behave are derived from observing how they do behave. However, I make every attempt to avoid turning the past into the present by uncritically mapping modern behaviours directly onto the archaeologically represented past. Ethnoarchaeological observations of material culture will be compared to the archaeological material culture and associated features (*e.g.* Agorsah 1990; Clarke 2001; Marshall 1987; Millerstrom 2006).

Archaeological studies of subsistence frequently apply the theoretical framework referred to as behavioural ecology (BE) or evolutionary ecology to understand zooarchaeological assemblages (*e.g.* Bird and O'Connell 2006; Burger *et al.* 2004; Butler 2001; Lupo 2006; Nagaoka 2002). BE and optimal foraging models seek to explain human and all animal behaviours based on the assumption that individuals relate to their environments in predictable ways, which maximize reproductive success. It is interesting to consider modern and contemporary examples of human eating and foodways in order to explore the usefulness of an optimal foraging approach across space and time. I have found that some of optimal foraging theory's underlying assumptions of 'rationality' and economics do not necessarily apply in modern situations in Fiji. I seek to understand ideas of success and value as Fijians define them. Finally, I suggest that research can be scientific and aim to interpret the past in emic terms, rather than making value judgments and assumptions of what sort of behaviour is rational versus illogical according to Western perspectives.

The Setting

Fiji's Lau Group was first inhabited about 3000 years ago by the Lapita peoples (Anderson and Clark 1999; Best 2002; Clark *et al.* 2001), the ancestors of the modern occupants of these islands. The Lapita culture complex is recognizable by distinctive dentate stamped pottery and associated stone and shell tool assemblages, the archaeological signature of the "seafaring pottery making farmers" who first inhabited Remote Oceania (Lilly 2006, 5). The Lau Group is an archipelago of 80 islands, 29 of which are currently inhabited, stretching north to south across 450km of ocean (Fig. 9.1). The main Fijian islands of Viti Levu and Vanua Levu are located about 200km west and 100km northwest of Lau, respectively. These volcanic and coralline limestone islands are located relatively close together and support extensive reef systems, rich in marine faunal resources.

My ethnographic data are the result of six months of participant observation on the island of Nayau and an additional four months of work on the nearby islands of Lakeba, Aiwa Levu, and Aiwa Lailai. Ethnoarchaeological work documented socioeconomic behaviours associated with rank and gender in qualitative and quantitative detail in order to facilitate comparisons between these ethnographic data and late prehistoric archaeological households in Fiji. Women's roles and associated material phenomena were recorded, especially food remains and marine resource debris, and were traced through two levels of analysis including ethnographic (modern lifeways) and ethnoarchaeological (modern consumption, deposition, and disposal).

Nayau (18.4km^2) is topographically and geologically varied, with limestone regions, large areas of volcanic soils, and bedrock outcrops. People on Nayau maintain extensive dryland and wetland agricultural crops, including taro (*Colocasia esculenta* and *Alocasia macrorrhiza*), cassava (*Manihot esculenta* and *M. dulcis*), sweet potato (*Ipomoea batatas*), and yam (*Dioscorea* spp.). These starch foods supplement inshore fish and other marine-associated resources such as shellfish, turtles, seaweeds, and occasionally larger bony fish caught offshore. Domestic animals contribute a smaller portion to the diet than bony fish. People typically eat pig or chicken once a week and during special occasions. Cows, a European introduction, are reserved for weddings and funerals.

Ethnographic Observations and Interviews

Ethnographic data indicate that four primary factors affect or determine decision-making and planning prior to and during inshore fishing expeditions (although many factors contribute, these general analytical groupings account for the most common situations). Fishing techniques differ according to: 1) natural factors; 2) the technologies accessible; 3) the number of people in the fishing party; and 4) the fish targeted (based on personal preference). In turn, these decisions work together to determine the make-up of any given catch, and ultimately what fish are consumed. These factors probably relate to modern and pre-contact subsistence situations in many island and coastal contexts. I describe each point in detail below.

Natural Factors

The first group of issues affecting decision-making is natural factors, including the weather, moon phase, tides, water depth, currents, and local environmental variations. For example, gathering particular species of molluscs and certain kinds of fishing (such as mullet fishing with large nets) increase during periods of daytime low tides and at new or full moons. Importantly, inshore and offshore areas around each village on a given island are known to be associated with different types of fishes. That is, each area is said to have better access to particular fishes depending on the reef morphology, substrate, and water depth. Because most fishing grounds are well known and sometimes have restricted access according to kinship, people generally are not secretive about fishing knowledge and readily share information.

Not surprisingly, people prefer to do inshore fishing at low tide, during daytime. Night fishing is relatively uncommon in central Lau, being practiced for specific occasions and with the intention of collecting particular fishes. On Nayau, men will occasionally use a flashlight and a spear inshore at night, targeting large carnivorous fishes. People with motorboats and long nets target mullet during the full moon.

Technology

Second, the technologies accessible to the fisherperson determine where, when, and how people may fish as well as what types of fishes are caught. Table 9.1 lists the Lauan names for different kinds of fishing gear and methods that are used on Nayau and Lakeba. The most common forms of fishing involve the use of nets in the inshore area. There are also a number of gender-specific terms associated with fishing technologies, such as the term for male vs. female fishing spears (*moto ni cocoka* and *moto ni nunu*). Ownership of particular items including fishing canoes, fibreglass boats with outboard motors, and nets are all increasingly based on a person's access to cash. At present, only relatively high ranked families or people who have relatives working off-island in lucrative jobs can afford imported items. The highest ranked male member of a household usually claims ownership of these goods. Micro-fibre nets greater than 20m long are the most sought after fishing technology, but these are expensive, heavy to carry, and require constant maintenance; thus only a few men own large nets in each village. For example, on Nayau, which is inhabited by about 450 people in three villages, there are only five micro-fibre nets suitable for large group fishing expeditions. Natural fibre nets have been out of use for about 50 years. In antiquity ownership of nets and fishing boats was contingent on rank and ability to harness resources to make or commission these items (Hocart 1929; Thompson 1940).

Today in Lau, adult women between the ages of 16 and 60 are the primary producers of inshore marine resources. Unmarried men and adolescents may join the women inshore, or groups comprised of two to twelve men occasionally go out together with spears, nets, and motor boats on the inshore or off-shore reef edge. Men of all ages fish on the outer reef edge with a mask and snorkel and a multi-pointed spear, a man's spear with four or more points; this is usually done alone rather than in a group. It is uncommon for families, including parents and children, to go fishing together. However, a husband and wife may

English Name	Lauan Term	Description
Fishing	*Qoli*	Any type of fishing
Fishing net	*Lawa*	General net
Mullet net	*Lawa toni*	"Dip net"; mesh usually 4cm mesh
Net to set	*Lawa toka*	Net that is set and left; "Placed net"
Crab net	*Lawa sua*	"Hobble along net"
Fine mesh net	*Lawa ni lovulovu*	Small gauge net for small fish
Fishing line	*Wa ni siwa*	Refers to microfiber or natural fiber line; *siwa* is also a verb, "to catch fish"
Fishhook	*Bati ni siwa*	"Bati" literally means "tooth"
Spear	*Moto*	General term
Man's fishing spear	*Moto ni cocoka*	Long spear with multiple points; cocoka is a verb, "to spear fish"
Woman's fishing spear	*Moto ni nunu*	Spear with one point; nunu is a verb, "to dive"
Charcoal sunscreen	*Loaloa*	Women rub charcoal on their faces to prevent sunburn
Group fishing	*Sabisabi*	Fishing in a group with nets, people chase fish into the net, swimming, kicking, throwing rocks, and slapping the water
Hand fishing	*Popono*	Fishing for small fish using hands only; children often do this
Mullet fishing	*Kadakada*	Fishing for mullet using a net
	Yavirau	Method used when many people participate; string is wrapped around coconut fronds and fish are chased into it and scooped up

Table 9.1 Fishing methods and gear used on Nayau and Lakeba

occasionally fish inshore, using spears and a small net. The frequency that women go out to the reef varies by individual, some women go everyday, weather permitting, and others fish two or three days a week, providing just enough to supply their family with a few fish each day and little excess.

The method of fishing in groups is referred to as *sabisabi*, and is the most common mode of group fishing. Microfibre nets are used; these are typically a single panel, 45m long, with 6cm or 10.5cm stretch mesh. *Sabisabi* requires one or two people to set the net and about three to six additional people that assemble some distance away (preferably with numerous coral heads in between the net and the swimmers). After the net is set, the swimmers spread out and begin to splash, kick, throw rocks, slap the water, and generally thrash around, scaring the fish into the net. When the fish hit the net the swimmers move along it, killing the fishes and tying them onto a stringer or putting them into a basket. All of the fish that hit the net are taken, regardless of size or type. Only on very rare occasions do people throw small fish back (<10cm total length).

Fishing Party Composition

Third, the number of people in the fishing party determines the areas exploited and the taxa targeted. The party's makeup, including the gender and rank of the members, determines how the catch is divided up and how many people receive portions. When a group brings their catch on shore they immediately begin to distribute it. Depending on the composition of the group and their preferences, the fish may be scaled and gutted on the beach, or simply taken away by the members of the fishing party to their respective households. As a result, the archaeological imprints of fish processing in domestic spaces are often unsubstantiated. Moreover, when fish are gutted and scaled on the beach or on the reef at low tide, this is rarely done in the same place; that is, there is no central place for handling and processing.

Regardless of whether the owner of the boat or net that were used is present, he will receive the first and largest portion of the catch (boats and nets are typically owned by men); this includes the biggest or most highly prized fish, such as a jack, mullet or parrotfish. After the owner(s) of the technologies are rewarded, a pile of fish is created for each of the participants, regardless of an individual's role in the expedition. If a husband and wife are both present, their pile will have approximately twice as many fish as the other piles. Lauans do not attempt to determine the relative contributions of the members of the fishing party when dividing up the catch. The person or people distributing the fish make every attempt to make each pile comparable or as they say, "equal." However, the piles are organized according to rank and age. The oldest, highest ranked individual present has a pile located next to that of the owner of the technologies. If only women are present, then the highest ranked woman will receive the first and most desirable division.

Unless the catch consists entirely of a single type of fish, such as mullet or parrotfish, there is wide variation in the

Lauan Name	Scientific Name	English Name	N
Kawakawa	Cephalopholis spp. and E. merra	Grouper	8
Mulu or Nuqa	Siganus sp.	Rabbitfish	7
Kanace	Mugilidae	Mullet	5
Tabacee	Acanthurus triostegus	Convict tang	3
Boosee	Scarus sp.	Parrotfish	3
Saqa	Caranx ignobilis	Giant trevally	3
Ika	Any fish	Fish	2
Vatui	Parupeneus cyclostomus	Goldsaddled goatfish	2
Kabajia	Lethrinus sp.	Emperorfish	2
Tabacee ni Toga	Acanthurus guttatus	Whitespotted surgeonfish	1
Saqa	Caranx melampygus	Bluefin trevally	1
Sevaseva	Plectorhinchus sp.	Sweetlips	1
Matu	Gerres sp.	Silver biddy	1
Jivijivi	Chaetodon sp.	Butterflyfish	1
Total			40

Table 9.2 Favourite fishes as indicated in interviews on Nayau. N = number of times fish cited as favorite.

sizes and taxa of fish collected. Heterogeneous catches are divided beginning with the choicest fish. Favoured fishes are distributed into the piles of the highest ranked persons (according to age, rank and gender) first. A fishing party with an excess of 'help' will receive a much smaller return on a successful expedition since every member is rewarded with a portion of the catch.

Individual and Group Preference

Fourth and perhaps most importantly, personal preference affects where people fish, what fish are targeted, the gear selected, and the fish that people ultimately consume. By conducting interviews with Lauans, I discovered some of the particulars of personal preference. Certain people love to eat small fish, while others, generally men, claim to prefer large fish. A sample of my data is presented in Table 9.2.

Interview data indicate that the majority of survey respondents on Nayau favoured inshore reef species that are typically small (<50cm in total length or TL). The single most sought after fish among women is the *kawakawa* (small groupers including *Epinephelus merra* and *Cephalopholis* sp., the honeycomb and hind groupers). *E. merra* in Fiji measure 20cm (TL) on average, while *Cephalopholis* averages 24–40cm in TL (Froese and Pauly 2006). People provided reasons for personal preferences including: colour, shape, degree of "sweetness," the size of the fish's eyes, and the colour of the fish's lips, among other things.

When I fished with groups of 8–10 people, twice over the course of one month the fishers decided to target schools of small Convict tangs (*Acanthurus triostegus*). These fish feed on filamentous algae in large aggregations, and reach an average length of 16cm, according to Fishbase (Froese and Pauly 2006), Myers (1991), and my own observations from Lau. These data indicate that preference is determined by characteristics other than the size of a fish.

In addition to the data collected during structured interviews, I frequently questioned people about what foods they preferred and why. When respondents were asked why they preferred certain fish to others they often referred to the size of the fish's head and eyes. These body parts are highly prized from both fish and mammals. One informant said that she liked emperor fishes (Lethrinidae) better than other fishes because, "It has red lips, big eyes, and sweet flesh." Another woman claimed that mullet was her favourite fish, "…because it is fat and has a big head." My principle Lauan collaborator, who sells fish to people all over central Lau, often targets mullet to sell; she explained that this is "…because it has firm flesh that lasts for a few days, and it has a big head and big eyes." All of the women interviewed claimed to prefer fish to all other types of meat including chicken, pig, and cow.

Household Consumption of Fish

Ethnographic information also provides a framework for understanding what a typical Lauan family consumes over a given period of time. I also aim to reconstruct the social context of the archaeological assemblages, including what portion of the diet this material represents and the

House #	# Adults	# Children	NISP	MNI	Weight (g)	N Types
1	2	2	594	20	43.7	14
2	4	2	1163	35	277.7	7
3	2	0	473	10	85.1	8
4	2	3	764	25	68.4	11
5a	2	2	788	22	74.9	9
5b	2	2	263	12	78.7	8
Total	14	11	4045	124	628.5	

Table 9.3 Totals for fish bone food remains collected from households on Nayau for one week. N Types = number of different species of fishes consumed.

importance of particular species to Lauan foodways. To follow-up the interviews on preference and to see how this affects what people consume, I asked five women in one village to save the fish bones from their family meals for one week, including all bony body parts. This sample includes the bones of fishes that were consumed by the entire household. The bones were collected after each meal, placed in sealed plastic containers and then analyzed. The five participants also kept running lists of the fishes they ate. I collected these lists and associated fish remains and then quantified the bones using the same methods that I apply to zooarchaeological samples (Table 9.3). I kept each sample separate to enable comparisons between the household's weekly fish consumption and to provide another example for inter-household comparisons. It should be noted that the bone samples I collected from each household were highly fragmentary and incomplete, representing only a fraction of the fish that were consumed by each household.

From the six samples, a total NISP of 4,045 fish bones were collected, weighing 628.5g (without any flesh attached and after drying). Total calculated MNI from this sample is 124. The household referred to as "number 5" collected remains for two weeks (labelled 5a and 5b in this Table). The five households exhibited great variation in the frequency of fishes consumed; NISP ranges from 263 to greater than 1,000, and calculated MNI varies from 10 to 35. The weight of the recovered fish remains is also highly variable. Household 2, with the largest sample of fish bones, yielded 277g of fish bones, while Household 1 produced only 43g.

The condition of the collected bones was uniform throughout the sample. All the fishes were prepared by boiling and no bones were burned. The sample consisted primarily of vertebrae and contained very few cranial elements, despite instructions to collect the entire skeletons of the consumed fishes. The participants explained that some of the fish heads were eaten or "disappeared" during their meals. On other occasions when I have asked people on Nayau to save all of the fish bones from their meals, they most commonly give me only the vertebrae and tail but not the head. I attribute the low frequency of cranial elements in the sample to the Fijian habit of dismantling

Taxon	NISP	g	MNI
Acanthurus triostegus	707	56.9	36
Terapon jarbua	479	21.7	12
Mullidae	409	22.6	17
Diodon sp.	386	85.5	4
Siganus sp.	235	31.8	9
Mugil spp.	220	144.5	5
Scarus sp.	219	93	9
Acanthurus spp.	160	17.4	7
Acanthuridae	108	18	0
Holocentridae	84	4.3	4
Muraenidae	79	13.4	1
Teraponidae	67	1.5	0
Acanthurus guttatus	45	15	2
Chaetodon sp.	43	4.3	2
Scaridae	33	1.1	0
Serranidae	28	2.5	3
Parupeneus sp.	24	2.9	1
Caranx sp.	22	13	1
Balistidae	21	10.1	3
Perciformes	19	2.4	0
Unidentified fragments	658	66.6	0
Total	4045	628.5	116

Table 9.4 Fish taxa consumed in sample households over a one-week period

the fish head, consuming all parts of it, and often biting or breaking it to release as much juice as possible. Pieces of fish crania are sometimes swallowed, or they may be spit out or thrown to the household dogs immediately.

The sample from five households includes 16 varieties of fishes, representing 13 families that were eaten during a week (Table 9.4). All of the households consumed both

Convict tangs (*Acanthrurus triostegus*) and parrotfish (Scaridae and/or *Scarus* spp.). Goatfish (Mullidae and/or *Parupeneus* spp.) were present in five of the six samples. By NISP and MNI surgeonfish (Acanthuridae, *Acanthurus guttatus*, *Acanthurus* spp., and *Acanthurus triostegus*) were the most frequently consumed taxa, followed by goatfish (Mullidae and *Parupeneus* sp.). By weight surgeonfish, mullet, porcupine fish, and parrotfish (Scaridae and *Scarus* sp.) were the dominant taxa. All of the aforementioned fishes are relatively small inshore reef taxa.

In terms of social status, Household 5, the home of a village headman (who functions like a Mayor) and his wife, is the highest ranked family that contributed fish bones to the sample. This family of two adults and two children (ages two and three) consumed a relatively moderate amount of fish in the first week's sample, ranking third by MNI, weight, and the number of different fish types consumed. This same household consumed considerably less fish in the second week by NISP and MNI. Importantly, the adults in this house eat far more than their young children.

Household 2 consumed the most fish by all measures, except in terms of the diversity of fishes eaten; by NISP, MNI and weight fish bones from this house greatly outnumber the other sample assemblages. Household 2 consumed 12 Convict tangs over the week. Six people, including four adults and two children (ages thirteen and six) live in this household, which appears to explain the relatively large amount of fish bones and high MNI in the sample. Household 3, occupied by only two adults, consumed a total MNI of 10 fish, yet yielded the second most abundant sample by weight.

Interpretations

This ethnoarchaeological study based on participant observation and interviews has produced information that illuminates Lauan decision-making about fishing and eating. Moreover, the data assist in understanding and interpreting the material consequences of behaviours and Lauan motivations. I have identified five critical points resulting from this research:

1. Lauan behaviour relating to fishing and foodways is the result of a number of interrelated natural and social issues including: the environment, access to technologies, group dynamics, hierarchy, and food preference.
2. Generalized collection strategies are the most common form of fishing (*sabisabi*). Fisherpeople (especially women) most often fish in groups in the inshore area and use nets. The resulting catch is very diverse but may include some species that are specifically targeted (for example, parrotfish, mullet, and Convict tangs).
3. Women between the ages of 16 and 60 are the primary producers of inshore resources.
4. Fish are a highly valued type of meat and inshore fishes are the most commonly consumed form of animal protein.
5. Fish preference is not based on size. Lauans consider characteristics other than the body size of a fish to determine its value, such as the size of the eyes. Relatively small (<50cm TL) inshore species, especially the honeycomb and hind groupers, are sought after.

By tracking household consumption I determined the amount of fish that are typically eaten in a week and the characteristics of the resulting faunal assemblage after consumption as well as the taphonomic factors impacting the material remains. Over a week households composed of 2–6 people (2–4 adults and 0–3 children) consumed between 263–1,163 NISP, 10–35 MNI, and 43–277g of fish. People ate between 7 and 14 different types of fish during the study period. This exercise also produced data on the condition of bones that might be included in a deposited assemblage. All of the fish were boiled and none of the bones were burned; all of the fish were highly fragmentary and the sample lacked any complete skeletons of any individuals. Eating practices resulted in fish assemblages that lacked most head elements. In fact, 78% of the material in the assemblage is comprised of vertebrae. The fish bones that remained after eating were far less copious than those associated with the actual consumed fish.

Correlations were found between the amount of fish consumed in the one-week sample and the number of people living in each household as well as the rank of its occupants. Household 5, the home of the highest ranked family produced a moderate amount of fish remains, however the two adults consumed almost all of the food since the children, ages 2 and 3 ate very little. The occupants of Household 5 also ate more chicken and pig during the week, which comprised more of the animal protein for them than in the other households where people ate inshore fish almost exclusively during the weekdays. Households other than number 5 supplemented their fish with a much smaller amount of chicken and/or pig. In addition, Household 2 has the largest number of occupants and consumed the largest quantity of fish by all measures.

The diversity of fishes present in the household samples does not correspond directly with family size or rank. No obvious link was found between the sizes of the reef fish in the assemblages and a household's rank. More data and a longitudinal perspective are needed to adequately evaluate these issues.

An obvious pattern from both the written food logs and the fish remains is the common consumption of inshore fish species. The preferences of the household occupants appear to have some affect on the composition of the fish assemblages and the taxa consumed each week. In each case the female head of the household who collected the fish remains explained to me why they ate particular types of fishes. All of the women commented that they like many types, or "all reef fish." However, when fish are collected on group fishing trips people will sometimes trade fish or ask for a specific favourite type or types. When women go out fishing alone they may look for particular kinds of fish, but they typically capture whatever fish they encounter.

While I recognize that this sample is small, it provides a foundation for future research, which will focus on expanding the sample size and documenting consumption over longer time periods. Long-term studies of modern fish collection and consumption should support or refute the aforementioned patterns concerning family and size and rank. Specific questions that will be examined with future research include: 1) Do faunal assemblages from modern contexts represent both household food preferences and the local availability of animal resources (a generalized collection strategy)? 2) Do people in high ranked households consistently consume more pig and chicken and less reef fish that other households? 3) Do high ranked people commonly eat larger fish than lower ranked people? 4) Is the collection and consumption of relatively small inshore reef species a long-term pattern that is practiced throughout the year?

Conclusions

A recent publication in the *Journal of Archaeological Research* equates "ethnoarchaeology" with the terms "site formation processes" and "taphonomy," in the catch-all category of "studies that link material consequences of behaviour with human behaviour itself" (Bird and O'Connell 2006, 144). Behavioural ecology (BE) has been widely applied as a means to make informed decisions about multiple possible readings of behavioural patterns associated with the generation of the archaeological record (Bird and O'Connell 2006; Shennan 2002). This approach is also seen as a means to guide speculation or interpretations about past behaviours that may not be represented archaeologically.

Many common Lauan resource collection, production, and consumption practices appear to be non-optimal. That is, these behaviours go against predictable foraging strategies with the best return for a given amount of effort. I argue that Lauan behaviours are better understood in the context of Fijian definitions for success and value as well as the local environment and indigenous economic systems.

A problem with the applicability of foraging models to inshore vertebrate collection is that most of the assumptions are founded on logical notions of collecting and processing shellfish and terrestrial game, which generally require more handling than reef vertebrates. The concept of optimal foraging "…is not an especially useful one when applied to Pacific Island fishing…" (Leach and Davidson 2000, 416) for three reasons. First, fish are abundant throughout the Pacific and they are relatively easy to catch in mass using net technologies that the Pacific Islanders have employed for millennia (Jones *et al.* 2007; Kirch and Green 2001). Second, the diets of Pacific Islanders included fish, domestic animals (dogs, pigs, and chickens) and other sources of energy, especially tubers. Pacific Islanders in Fiji and Polynesia have generally had little difficulty collecting enough bony fish and other marine resources to support themselves. Third, optimal foraging theory assumes that people seek out and prefer large bodied animals over small bodied animals and, in the case of fishing, people will always collect large fish over small fish. Both the archaeological and ethnoarchaeological data from Lau refute this notion (Jones 2007; Jones *et al.* 2007).

Ethnoarchaeological observations remind us that as archaeologists we can benefit by understanding local conceptions of valued foods and ideas of what a successful hunting, fishing, or gathering expedition is in order to predict what will be collected and to better interpret zooarchaeological assemblages. Local concepts of value do not always mesh with Western ideas about what is logical or economical. Ethnoarchaeology can form the basis for models about practice that are grounded in reality rather than just theory. This work is fundamentally comparative and does not seek to find laws but to enhance understanding. This approach involves interpretive discovery that examines the complexity of ethnographic records and human behaviour, observing the social contexts in which material is collected, used, and discarded, with a focus on social issues and processes. I hope that this paper will not be viewed as a contentious commentary on a particular theoretical approach. Rather, my intention is to point out that a historically informed archaeology, which is sensitive to context can also be scientific. Moreover, regardless of one's theoretical focus, if data are generated using standardized methods and are included in publications, it may benefit many researchers.

Acknowledgments

Many thanks to Umberto Albarella for inviting me to contribute to this session and the associated volume. I am indebted to the people who assisted with and made this research possible. In particular I thank David Steadman, Sepeti Matararaba, Sean Connaughton, Rusil Colati, and Sepesa Colati. I gratefully acknowledge the late *Na Gone Turaga Na Tui Lau, Tui Nayau Ka Sau ni Vanua ko Lau*, The Right Honourable Ratu Sir Kamisese Mara for allowing me to work in Lau. I also thank the Fiji Museum for its important assistance. The people of Nayau and Lakeba provided me with a wealth of local knowledge, *vinaka vakalevu*. Part of this research was funded by a UAB-NSF ADVANCE grant to the author.

References

Agorsah, E. K. (1990) Ethnoarchaeology: The Search for a Self-Corrective Approach to the Study of Past Human Behaviour. *The African Archaeological Review* 8, 189–208.

Anderson, A. and Clark, G. (1999) The age of Lapita settlement in Fiji. *Archaeology in Oceania* 34, 31–39.

Best, S. B. (2002*) Lapita: A View From the East.* New Zealand Archaeological Monograph 24. New Zealand Archaeological Association Publications. Auckland, Auckland Museum.

Bird, D. W. and O'Connell, J. F. (2006) Behavioral Ecology and Archaeology. *Journal of Archaeological Research* 14(2), 143–188.

Butler, V. (2001) Changing Fish Use on Mangaia, Southern Cook Islands: Resource Depression and the Prey Choice Model. *International Journal of Osteoarchaeology* 11, 88–100.

Burger, O., Hamilton, M. J. and Walker, R. (2004) The Prey and patch model: optimal handling of resources with diminishing returns. *Journal of Archaeological Science* 32, 1147–1158.

Clark, G., Anderson, A. and Matararaba, S. (2001) The Lapita site at Votua, northern Lau Islands, Fiji. *Archaeology in Oceania* 36, 134–145.

Clarke, M. J. (2001) Akha feasting: An ethnoarchaeological perspective. In M. Dietler and B. Hayden (eds.) *Feasts: Archaeological and Ethnographic Perspectives on Food, Politics, and Power*, 144–167. Washington, DC, Smithsonian Institution Press.

Froese, R. and Pauly, D. (eds.) (2006) *Fishbase: A Global Information System on Fishes*. Accessed December 2006, http://www.fishbase.org.

Hocart, A. M. (1929) *The Lau Islands, Fiji*. B. P. Bishop Museum Bulletin 62. Honolulu, Bishop Museum Press.

Jones, S. (2007) Human Impacts on Ancient Marine Environments of Fiji's Lau Group: Current Ethnoarchaeological and Archaeological Research. *The Journal of Island and Coastal Archaeology* 2(2), 1–6.

Jones, S., Steadman, D. W. and O'Day, P. (2007) Archaeological investigations on the small islands of Aiwa Levu and Aiwa Lailai, Lau Group, Fiji. *Journal of Island and Coastal Archaeology* 2, 72–98.

Kirch, P. V. and Green, R. C. (2001) *Hawaiki, Ancestral Polynesia: An Essay in Historical Anthropology*. Cambridge, Cambridge University Press.

Leach, F. and Davidson, J. (2000) Fishing: a neglected aspect of Oceanic economy. In A. Anderson and T. Murray (eds.) *Australian Archaeologist: Collected Papers in Honour of Jim Allen*, 412–426. Canberra, Coombs Academic Publishing, The Australian National University.

Lilly, I. (2006) *Archaeology of Oceania: Australia, and the Pacific Islands*. Blackwell Studies in Global Archaeology. Oxford, Blackwell Publishing.

Lupo, K. D. (2006) What explains the carcass field processing and transport decisions of contemporary hunter-gatherers? Measures of economic anatomy and zooarchaeological skeletal part representation. *Journal of Archaeological Method and Theory* 13(1), 19–66.

Marshall, Y. (1987) Maori Mass Capture of Freshwater Eels: an Ethnoarchaeological Reconstruction of Prehistoric Subsistence and Social Behavior. *New Zealand Journal of Archaeology* 9, 55–79.

Millerstrom, S. (2006) Ritual and domestic architecture, sacred places and images: Archaeology in the Marquesas Archipelago, French Polynesia. In I. Lilly (ed.) *Archaeology of Oceania: Australia, and the Pacific Islands*. Blackwell Studies in Global Archaeology. Oxford, Blackwell Publishing.

Myers, R. F. (1991) *Micronesian Reef Fishes: A Practical Guide to the Identification of the Coral Reef Fishes of the Tropical Central and Western Pacific*. 2nd Edition. Guam, Coral Graphics.

Nagaoka, L. (2002) Explaining subsistence change in Southern New Zealand using Foraging Theory models. *World Archaeology* 34, 84–102.

Schmidt, P. R. (1997) *Iron Technology in East Africa: Symbolism, Science, and Archaeology*. Bloomington, Indiana University Press.

Shennan, S. (2002) Archaeology and Evolutionary Ecology. *World Archaeology* 34, 1–5.

Thompson, L. M. (1940) *Southern Lau, Fiji: An Ethnography*. B. P. Bishop Museum Bulletin 162. Honolulu, Bishop Museum Press.

10. Ethnozooarchaeology of the Mani (Orang Asli) of Trang Province, Southern Thailand: A preliminary result of faunal analysis at Sakai Cave

Hitomi Hongo and Prasit Auetrakulvit

The Mani hunter-gatherer groups inhabit the tropical rainforest of southern Thailand and northern Malaysia. This paper presents results from analysis of animal bone remains collected during an ethnoarchaeological investigation of Sakai Cave in Trang Province (southern Thailand) during an excavation by Silpakorn University and University of Tübingen. Faunal remains reported here were left by modern Mani groups who have used the cave as one of their campsites. The material accumulated on the floor of the cave probably in the last 50 years, but perhaps in the last 100 years. The Mani specialize in hunting arboreal animals using a blowpipe. The results of analysis suggest that primates make up the majority of faunal remains left at the Mani campsite. Langurs are the most favored game, followed by gibbons and macaques. A wide range of mammals, including giant squirrels and various species of Viverridae, birds, and fish are also exploited. Among terrestrial fauna, turtles and lizards are most abundant. Butchery of animals is characterized by cutting off both epiphyses of long bones, which was probably a practice started by the recent introduction of metal tools.

Key words: The Mani, Thailand, Sakai Cave, blowpipe, primates

Introduction

This paper presents preliminary results of the analysis of faunal remains excavated from Sakai Cave in Trang Province in Southern Thailand. The cave had until recently been used by the Mani, a group of nomadic hunter-foragers, as one of their camp sites. The animal bone remains recovered from the site are considered to have been accumulated by the Mani groups who periodically visited and stayed in the cave. The archaeologists believe that the excavated animal bones were accumulated recently, perhaps only within the last one hundred years (H. P. Uerpmann pers. comm. 2005). The traditional subsistence of the Mani has been affected by various political and economical factors; fighting between government forces and anti-government communist guerrillas based in the forest in the 1970's destroyed their habitat (Pookajorn 1991, 262), while commercial deforestation intensified (Nobuta 2005). Not only did game in the forest decrease, but many of the wild animals that the Mani traditionally hunted are now on the list of endangered species and their hunting is banned. The Mani are subject to pressure to settle down. They increasingly spend their time working as agricultural labourers or living under social welfare at the 'resettlement' villages constructed by the Thai government (Hamilton 2002). Therefore, the faunal remains from the Sakai Cave provide us with important information about the exploitation of wild animals by the Mani, such as the range of animal species hunted and their relative frequencies, before their traditional way of life was drastically changed.

The Mani

The homeland of the Mani is the tropical rainforest of southern Thailand and Northern Malaysia. The Mani bands on both sides of the Thai–Malay border are linked by kinship and move across the border (Hamilton 2002). The groups in Malaysia are called Orang Asli. The Mani in Thailand are also called "Sakai" by the local Thai, from which the name of the cave was derived. However, the name carries a negative connotation meaning 'savage' or 'slave.' They call themselves Mani or Maniq, which means 'human being' in the Kensiw language (Hamilton 2002, 82). In appearance, the Mani have Negrito characteristics: their complexion is dark; their hair is curly; and their stature is short, although it has been reported recently that the younger Mani people, whose diet has become more similar to the sedentary farmers, have become much taller (H. Müller-Beck pers. comm.).

Fig. 10.1 Map of Thailand and the location of Trang Province

The Mani are seen in Thai culture as magical beings of the forest and allies of the royal court. This notion originated from the poem written by King Chulalongkorn, who took a Mani orphan to the court, in the beginning of the 20th century (Porath 2002). In the mid-1990's, there were about 140 Mani in Thailand, which is unchanged since the observation by H. Bernatzik in the 1920's (Bernatzik 1938; Albrecht and Moser 1998). There are 3 bands in Trang and Pattharung Provinces, another small group in Satun Province, and two groups in Yala Province (Hamilton 2002, 84–85)(Fig 10.1). One of the groups in Yala is settled in a village, and is also incorporated into the tourist industry and gives performances in the cultural shows.

A group of Manis consists of four to forty people (Pookajorn 1996, 447), and usually includes a married couple, children, in-laws and relatives. The traditional Mani camp is a simple shelter with a roof made of banana leaves (Fig. 10.2). They also use caves or rock shelters as campsites. Simple beds are made with split bamboos and

Fig. 10.2 Shelter used by the Manis (photo by S. Pookajorn)

tree bark, with hearths near each bed (Fig. 10.3). There are also huts made of bamboo, but this type of housing began to be used after the Thai government's pressure on

Fig. 10.3 Inside of a rock shelter used by the Manis (photo by S. Pookajorn)

Fig. 10.4 Sakai Cave before excavation in 1991 (photo by S. Pookajorn)

the Mani to settle down. They usually stay at one camp for 10–15 days, until the colour of the leaf roofing fades (Pookajorn 1996, 454). They also move when food around the camp, mainly wild yam potatoes, becomes scarce, or in order to obtain bamboo for making blowpipes to be used in hunting. There are other reasons for moving, for example, when they have bad dreams, become sick, or when someone dies (Pookajorn 1996, 455). When a few Neolithic graves were found in the excavation of Sakai Cave in 1991 (Pookajorn 1996), the Mani group that was staying in the cave moved away and stopped using the cave as a camp (H. Müller-Beck pers. comm.). This migration pattern seems to be changing recently, which might be at least partly due to less favourable living conditions as a result of deforestation. German researchers observed in 1995 and 1996 that the size of the bands was becoming smaller (Albrecht and Moser 1998, 162), and that moves were more frequent. One of the groups visited nine camps in 30 days in December 1995 and January 1996, with each camp being used only for one to four days. The researchers consider food shortage to be the cause of the frequent movements (Albrecht and Moser 1998, 191).

The most important food items are yam potatoes and other plant roots. Thonghom lists 10 kinds of yams, 10 kinds of leaves and bamboo shoots, and 17 kinds of fruits as plant food items used by the Mani, and notes that many other vegetables and fruits are also used (Thonghom 1995). Honey is collected and sold to sedentary farmers and is an important source of cash income.

The Mani use blowpipes (*Bolau*) made of bamboo for hunting. The blowpipe is about 1.5–2 meters long and about 2 centimeters in inner diameter. Poison extracted from the resin of *Antiaris toxicaria* is applied to the tip of the darts (*Bila*). There is a debate on the antiquity of blowpipes (Rabett and Piper 2006; King 1995; Sloan 1975; Oppenheimer 1998), and it has been suggested that slings or bows and arrows might have been used in earlier times. As discussed below, animals in trees are main targets of hunting. Meat is, however, not the main food of the Mani, and plant food is far more important in their diet (Thonghom 1995).

Sakai Cave

Sakai Cave (Fig. 10.4) is located in Palian District, Trang Province, in southern Thailand (Fig. 10.1). The cave was used as a campsite by the Mani until recently, but there are also prehistoric deposits belonging to the Pre-Neolithic and Neolithic under the cave floor. Archaeological investigations of Sakai Cave were first carried out in 1991 under the direction of Professor Surin Pookajorn at the Department of Archaeology of Silpakorn University in Bangkok, as part of the 'Hoabinian Research Project in Thailand' (Pookajorn 1991; 1994). The project's goal was to investigate human activities and environmental changes during the period of the Pleistocene–Holocene transition in this part of Southeast Asia, as well as to collect ethnological observations of the contemporary hunter-gatherers who inhabit the tropical rainforest of southern Thailand (Pookajorn 1991; 1996).

Three cultural layers were identified during the excavation of Sakai Cave. The oldest, Layer 1, is pre-Neolithic, and pebble tools similar to Hoabinian stone tools and hearths were found within it. Two Carbon 14 dates, 9,280 ±180 BP and 9,020 ±360 BP, were obtained from this layer (Pookajorn 1996, 430). A Neolithic deposit (Layer 2) was found immediately above the pre-Neolithic layer. Typical pottery found in Layer 2 was coarse ware decorated with cord marks. Based on polished stone axe typology, Layer 2 could be dated to before 1400 BC. Two burials, an adult female and a girl, were also found. The uppermost Layer 3 consisted of recent remains left by modern Mani groups who visit and stay in the cave at least once a year during the rainy season between April and December (Pookajorn 1996, 433). The thickness of this uppermost layer was about 50cm. The cave was still regularly used by the Mani at the time of the excavation. But when burials were found in Layer 2, the band moved away and never came back (Moser 1999; H. Müller-Beck pers. comm. 2004). Apparently the Mani had no knowledge of the Neolithic burials and avoided the cave once they knew that human skeletons were found there.

Starting from 1992, a joint ethnoarchaeological project

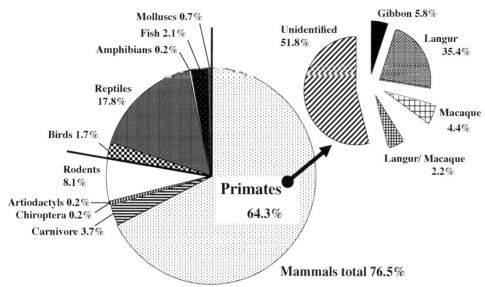

Fig. 10.5 Relative frequency of identified taxa at Sakai Cave (NISP=2575)

in southern Thailand was carried out by the University of Tübingen in Germany, and Silpakorn University in Bangkok, Thailand. The project's primary aim was to document the way of life of the Mani before it was drastically changed by outside pressure (Albrecht 1994; Albrecht and Moser 1995; 1998; Albrecht et al. 1993).

Faunal Remains and the Method of Analysis

The faunal material currently under analysis was collected from Sakai Cave during the 1997 field season. All the material belongs to Layer 3, the recent Mani remains accumulated probably within the last 50 years or so. Pieces of plastic and glass are often mixed in with the recovered material, which confirms this view.

A total of about 3500 fragments has been analysed so far, which is about 70% of the entire assemblage from the site. The faunal remains were identified using the modern skeletons of South East Asian mammals borrowed from the Naturkundemuseum in Stuttgart and the Department of Zoology in the University of Tübingen. In addition, comparative specimens of mammals, birds and reptiles in the archaeozoology laboratory of University of Tübingen were used in the analysis.

Previous Reports on the Faunal Remains from Sakai Cave

Animal bones excavated from Sakai Cave during archaeological field work in 1991 were identified and reported by Chaimanee, but the information is limited to a list of taxa and a minimum number of individuals (MNI) count based only on jaws and teeth (Chaimanee 1996). The material from the 1991 season was collected from all three layers at the site (pre-Neolithic, Neolithic, and recent Mani remains). The identified taxa include langur, macaque, hog badger, pig, giant squirrel, flying squirrel, porcupine, fish and turtle. Since the species MNI was in many cases only one, it is impossible to estimate the relative importance of each taxon. Within the limited data, langur, however, seems to have been the most abundant taxon in the assemblage. It is the only taxon whose MNI was greater than one, up to four in some layers.

Berke identified the faunal remains from the 1993 excavation season at Sakai Cave (Albrecht et al. 1993). More than 5000 fragments were recovered from layers dated between about 10,000 to 9,500 BP. Primates bones make up about 36% of the 423 identified specimens. Gibbons are the most abundant. More macaques than langurs were identified, but more than half of the primate bones were just classified as either langurs or gibbons. Therefore, langurs could actually have been more frequently hunted than macaques. Pangolin and land turtles are also important taxa; each makes up close to 15% of the identified specimens.

Result of Analysis of Faunal Remains from 1997 Season

Table 10.1 lists the animal taxa found in the faunal assemblage from the 1997 season. Based on the NISP, about two thirds (64%) of the identified specimens come from primates (Fig. 10.5). One small humerus was identified as that of a slow loris. Next in frequency are giant squirrels, which amount to 7.6% of the identified specimens. Viverrid bones comprise a little more than 2% of the identified specimens. Binturong is the most common species among the Viverridae. Bones of small squirrels and flying lemurs were also encountered in the assemblage. Thus the majority of hunted animals are arboreal in their nature, which is not surprising as the main hunting tool of the Mani is the blowpipe. Birds are sometimes hunted; hornbills seem to be the most common, but small Turdidae were also identified. Nocturnal species, such as binturong, are hunted without using blowpipes during the daytime,

Mammalia		
Dermoptera		
	Flying Lemur	*Cynocephalus variegatus*
Chiroptera		
	Unidentified bat	
Primates		
	Gibbon	*Hylobates* sp. cf. *lar*
	Langur	*Trachypithecus* sp. cf. *obscurus*
	Pig-tailed Macaque	*Macaca nemestrina*
	Slow Loris	*Nycticebus coucang*
Carnivora		
	Binturong	*Arctictis binturong*
	Masked Palm Civet	*Paguma larvata*
	Little Civet	*Viverricula indica*
	Hog Badger	*Arctonyx collaris*
	Dog	*Canis familiaris*
	Otter?	*Lutra* sp.?
Artiodactyla		
	Muntjac	*Muntiacus muntjak*
Rodentia		
	Giant Squirrel	*Ratufa bicolor*
	Squirrel	*Callosciurus* sp.
	Flying squirrel	*Hylopetes* sp.
	Porcupine?	*Hystrix brachyura/ Atherurus macrourus*
Aves		
	Hornbill	*Buceros bicornis*
	Starling?	Turdidae?
Reptilia		
	Varanus	*Varanus salvator*
	Soft shell turtle	*Manouria impressa*
Amphybia		
	Unidentified frog	
Osteichthyes		
	Cyprinidae	
Gastropoda		
		Brotia costula

Table 10.1 List of identified Taxa at Sakai Cave

when they are sleeping on their nests (pers. comm. from Mani hunters to P. Auetrakulvit). There are very few bones of terrestrial animals, though hog badger is sometimes hunted and a few bones of small deer were also found in the assemblage. Turtles, both land turtles and soft-shell turtles, are commonly exploited, as well as the giant lizard. They are easy to catch and are an important sources of food for the Mani.

Exploitation of Primates

The majority of primate bones are those of langurs, followed by gibbons (Fig. 10.5). Of the 1655 primate remains, 732 specimens were identified to genus. 77.6% of these were langur bones. Bones of gibbons and macaques amount to 12.7% and 9.7% respectively. Based on size and morphology, the macaques are identified as pig-tailed macaque (*Macaca nemestrina*). Since macaques move on

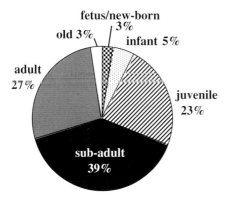

Fig. 10.6 Age of primates at Sakai Cave (n=337)

	number of specimens	%
Female	46	71.9
Male	18	28.1
total	64	

Table 10.2 Sex ratio of primates from Sakai Cave

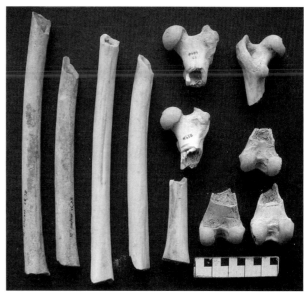

Fig. 10.7 Chopped femurs of primates from Sakai Cave

the ground as well as in the trees, langurs may therefore be better quarry in blowpipe hunting.

Age at death of the primates was estimated using the state of epiphyseal fusion and tooth eruption and wear, following Harvati (2000). The ageing of the primates is still preliminary, and all three species of primates are grouped together in the present analysis. Tooth eruption and wear as well as the state of epiphyseal fusion suggest that many of the primates in the assemblage are adult and subadult, but individuals of all age classes including infants are present in the sample (Fig 10.6).

More female primates are hunted than males. The female to male ratio is about 2.6 to 1 (Table 10.2). This is especially the case for gibbons – out of the 12 jaws, teeth, and pelves that could be sexed, 10 were females. Such a hunting strategy focusing on females, together with the indiscriminate hunting of young and juvenile individuals, undoubtedly resulted in high hunting pressure on the primate population. This is exacerbated because primates give birth to only one offspring every few years.

Butchery

In the case of primates and giant squirrels, all skeletal elements, including phalanges, caudal vertebrae, skulls and mandibles are present at the site. Carpals or tarsals and metapodials recovered from the same feature often articulate. Thus, primates were brought back to the cave complete and butchered on site. In the case of other, less important taxa, such as hog badger, binturong and flying lemur, not all of the skeletal elements are present at the site. Skulls and extremities are, however, often encountered, which suggests that these animals were also butchered on site.

A characteristic and uniform butchery method was

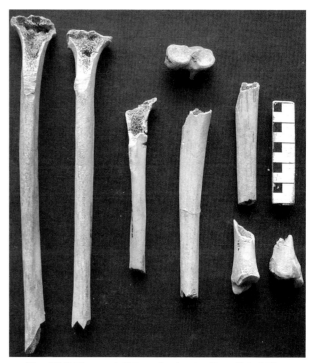

Fig. 10.8 Chopped tibiae of primates from Sakai Cave

observed on the bones of primates and giant squirrels. Both epiphyses of long bones were chopped off, probably to use the marrow (Figs. 10.7 and 10.8). This butchery technique was applied to all mammal bones irrespective of species. Even relatively small limb bones of squirrels and phalanges of primates were cut in the same way (Fig. 10.9). For the phalanges, in many cases the proximal part was chewed after the proximal end was cut off. In addition, the dorsal side of the proximal tibia of primates was cut off, probably to remove the patella. Such butchery practices are apparently only possible with the use of sharp metal knives, which is probably a rather recent phenomenon after

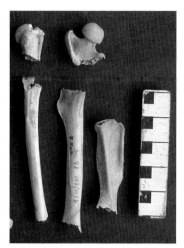

Fig. 10.9 Squirrel bones with chop marks

frequent contacts with sedentary farmers. Long bones of binturong are often found complete, although some of them bear cut marks. Binturongs are arboreal nocturnal animals, and one of the authors (Auetrakulvit) was told that the local people can easily catch them during the day while the animals are sleeping. The hunting method as well as butchery procedure for binturong might have been different from that for the primates. The number of binturong long bones in the sample is, however, too small to permit any concrete conclusions to be drawn.

Summary and Discussion

The faunal remains from Sakai Cave currently under analysis are considered as relatively recent material, accumulated by contemporary Mani groups who have used the cave on a regular basis. Even though the material was perhaps as little as 50 years old and not more than 100 years old, it reflects the traditional hunting practices of the Mani before their life and subsistence activities were drastically changed by modern political and economic pressures in the 1970s. The Mani hunt using blowpipes and poisonous darts, targeting arboreal animals, with the main source of animal protein being primates. Primate bones make up about two thirds of faunal remains. It is not surprising that such a hunting strategy has been seriously affected by the reduction of forest area in recent years. The majority of the primate bones from Sakai Cave are of langur. Preliminary observations of age and sex of the hunted primates suggest that the Mani hunters tend to take more females than males, and are not selective with regard to age classes. The Mani groups who used the cave exploited the animals available in the forest around the cave, and, as the modern Mani groups do, moved to another campsite when the animal and plant resources in the immediate surrounding of the cave were exhausted. This hunting practice was sustainable only when the Mani could move around vast areas of tropical rainforest, and when animal resources were abundant. One of the authors (Auetrakulvit) observed during the field work at Sakai Cave that the hunters were not successful in hunting primates, and in most cases came back from their hunting expeditions empty-handed.

The age profiles of gibbons, langurs, and macaques need to be compared in order to investigate whether a different hunting strategy was employed for each species. Since all primates in Thai forest are now protected, it is not possible to hunt them openly any more, although the hunting activities of the Mani may be left unchecked in most cases.

Hunted animals were brought back to the cave and butchered there. The characteristic chopping marks observed on the faunal remains from Sakai Cave were apparently made by using metal knives. This butchery method possibly reflects the practices employed even before the Mani obtained metal tools through contacts with sedentary farmers. Since butchery marks on the bones from Neolithic and pre-Neolithic layers at Sakai Cave have not been studied, it is unknown whether the chopping off of both ends of the long bones was a practice that can be traced back to prehistoric times. The study of butchery marks on faunal remains from the West Mouth of Niah Cave in Malaysia (*c.* 45,000 to 8,000 BP) reports that the hind leg was removed by chopping immediately beneath the femoral head. Cut marks are rather uncommon on the animal bone remains from Moh Khiew Cave II in Krabi Province in Thailand (*c.* 28,500 to 4,000 BP) (Auetrakulvit 2004). Many of the primate long bones from Moh Khiew Cave II were, however, broken near the epiphyses (Auetrakulvit 2004, Plate 4). These two examples may suggest a common butchery practice in Southeast Asia with a long tradition.

Even though the present study of faunal remains from Sakai Cave is still in its preliminary stages, the results should provide insights into the hunting strategy and butchery techniques of the Mani, whose traditional subsistence is on the verge of disappearance because of deforestation and political/economic pressures.

Acknowledgements

Funding was provided by the HOPE Project sponsored by the Japan Society for the Promotion of Science (Primate Research Institute, Kyoto University and Max Planck Institute for Anthropology). We would like to thank Professor H.-P. Uerpmann at University of Tübingen for giving us for the opportunity to study the animal bone remains from Sakai Cave. We are grateful to Professor H. Müller-Beck and Dr. G. Albrecht for generously providing us with valuable information on the excavation of Sakai Cave and ethnoarchaeological study of the Mani. We are especially thankful to Mr. Kurt Langguth and Students and colleagues at Archaeozoology Lab, U. of Tübingen for their help. The comparative skeletons of Southeast Asian animals used for the study were borrowed from the Zoology Department, University of Tübingen and Naturkundemuseum in Stuttgart. Dr. Doris Mörike at Naturkundemuseum provided her help in selecting and locating the skeletons.

References

Albrecht, G. (1994) Das Abri La Yuan Pueng- ein Siedlungsplatz der Mani in der Satun Provinz/Südthailand. *Ethnographisch-Archaologische Zeitschrift* 35, 199–207.

Albrecht, G. and Moser, J. (1995) Recent Mani Settlements in Satun Province, Southern Thailand: Report on the 1995 field campaign. University of Tübingen.

Albrecht, G. and Moser, J. (1998) Recent Mani Settlements in Satun Province, Southern Thailand. *Journal of the Siam Society* 86(1 & 2), 161–199.

Albrecht, G., Berke, H., Burger, D., Moser, J., Müller-Beck, H., Pookajorn, S., Rähle, W. and Urban, B. (1993) Sakai Cave, Trang Province – Southern Thailand: Report on the Field Work 1993. University of Tübingen.

Auetrakulvit, P. (2004) Faunes du Pléistocène final à l'Holocéne de Thaïlande: approche archéozoologique. Unpublished dissertation. University of Aix-Marseille I.

Bernatzik, H. A. (1938) Die Geister der gelben Blätter. Munich, F. Bruckmann.

Chaimanee, Y. (1996) Mammalian Fauna from Archaeological Context from Moh Khiew and Sakai Caves. In S. Pookajorn S. and staff, *Final Report of Excavations at Moh Khiew Cave, Krabi Province; Sakai Cave, Trang Province and Ethnoarchaeological Research of Hunter-Gatherer Group, Socall Mani or Sakai or Orang Asli at Trang Province*, 405–418. The Hoabinhian Research Project in Thailand, Volume 2: 1994. Department of Archaeology, Silpakorn University.

Hamilton, A. (2002) Tribal People on the Southern Thai Border: Internal Colonialism, Minorities, and the State. In G. Benjamin and C. Chou (eds.) *Tribal Communities in the Malay World: Historical Cultural and Social Perspectives*, 77–96. Singapore, Institute of Southeast Asian Studies.

Harvati, K. (2000) Dental Eruption Sequence Among Colobine Primates. *American Journal of Physical Anthropology* 112, 69–85.

King, V. (1995) Tropical Rainforests and Indigenous Peoples: Symbiosis and Exploitation. *Sarawak Museum Journal (New Series)* 69, 1–22.

Moser, J. (1999) Recent Cave Dwellings in Southeast Asia: Homes, Domiciles or Refuges? Explanation and Interpretation of Prehistoric Archaeological Structures. In L. R. Owen and M. Porr (eds.) *Ethno-Analogy and the Reconstruction of Prehistoric Artefact Use and Production*, 275–279. Tübingen, Mo Vince Verlag.

Nobuta, T. (2005) People who Reuse the Forest. Harvest of Durian Fruits by Orang Asli. In K. Ikeya (ed.) *Forest People in Tropical Asia*, 223–250. Tokyo, Jinbunshoin.

Oppenheimer, S. (1998) *Eden in the East: The Drowned Continent of Southeast Asia*. London, Weidenfeld and Nicolson.

Pookajorn, S. (1996) Human Activities and Environmental Changes during the Late Pleistocene to Middle Holocene in Southern Thailand and Southeast Asia. In L. G. Straus, B. V. Eriksen, J. M. Erlandson and D. R. Yesner (eds.) *Humans at the End of the Ice Age: The Archaeology of the Pleistocene-Holocene Transition*, 201–213. New York, Plenum Press.

Pookajorn, S. and staff (1991) *Preliminary Report of Excavations at Moh Khiew Cave, Krabi Province; Sakai Cave, Trang Province and Ethnoarchaeological Research of Hunter-Gatherer Group, Socall "Sakai" or "Semang" at Trang Province*. The Hoabinhian Research Project in Thailand, Volume 1; 1991. Department of Archaeology, Silpakorn University.

Pookajorn, S. and staff (1996) *Final Report of Excavations at Moh Khiew Cave, Krabi Province; Sakai Cave, Trang Province and Ethnoarchaeological Research of Hunter-Gatherer Group, Socall Mani or Sakai or Orang Asli at Trang Province*. The Hoabinhian Research Project in Thailand, Volume 2; 1994. Department of Archaeology, Silpakorn University.

Porath, N. (2002) Developing Indigenous Comunities into *Sakais*: South Thailand and Riau. In G. Benjamin and C. Chou (eds.) *Tribal Communities in the Malay World: Historical Cultural and Social Perspectives*, 96–118. Singapore, Institute of Southeast Asian Studies.

Rabett, R. J. and Piper, P. J. (2006) Eating your tools: early butchery and craft modification of primate bones in tropical Southeast Asia. Unpublished manuscript.

Sloan, C. (1975) Punan hunting methods. *Sarawak Museum Journal* 20 (New Series 40–41), 262–269.

Thongham, S. (1995) Sakai: Forest People. Provincial Administrative Organization, Trang Province. The Thanvela Co. Ltd.

Food Preparation and Consumption

11. An ethnoarchaeological study of marine coastal fish butchery in Pakistan

William R. Belcher

Archaeological fish remains allow a direct means to reconstruct aquatic resource use by past human populations. However, unlike mammalian fauna, little research has been done in reference to the study of ethnographic fish butchery practices that allow us to interpret the archaeological record. The research reported here presents a study of fish processing and consumption in complex village and urban settings of modern, coastal Pakistan. This ethnoarchaeological research focuses on the fishing villages of Buleji and Hawkes Bay and the large urban centre of Karachi. Significant differences occur in fish butchery and consumption for local village use as opposed to fishes prepared in large market places for commercial distribution of dried and fresh fish. Fish species and butchery 'style' differences in various urban market places appear to be related to the ethnic and socio-economic status of the neighbourhood. These general, idealized models, based on ethnographic observations, can aid in the understanding of fish resources in ancient complex societies. These models are a first step in providing a more quantitative understanding of urban-level fish butchery.

Keywords: Pakistan, fish butchery, Buleji region, market places

Introduction

Fish are an important component of the subsistence strategies of many past and present human societies (*e.g.* Stewart 1987; 1991; Stewart and Gifford-Gonzalez 1994; Fagan 2006). However, zooarchaeological research has tended to focus on reconstructing fishing strategies as well as taphonomic processes (*e.g.* Lubinski 1996; Butler 1993; Butler and Schroeder 1998; Richter 1986; Nicholson 1996; Willis *et al.* 2007) with less focus on butchery practices, unlike studies of terrestrial mammalian fauna (*e.g.* Lyman 1986; Lupo 2001; Bunn 1994; 2001; however, see B. Hoffman *et al.* [2000] or Zohar and Dayan [2001] for notable exceptions). Additionally, many archaeological, ethnographic, and experimental fishing studies have focused on hunter-gatherer-fisher societies or pastoral groups and not the use of fish in the more complex, cultural settings of sedentary village and urban centres. Most studies of more complex societies usually are dependent on literary or historical references (R. Hoffman 1994; de Jong 1994; Raychaudhuri 1980). However, within the last decade notable research has been conducted on historical fishing and its context within the preparation and distribution of European stock fish (*i.e.* Locker 2001). By using an ethnoarchaeological approach towards fish butchery and redistribution within village and urban settings, this paper examines the role of fish in these more complex societies (*e.g.* Colley 1990; Brinkhuizen 1994; Wendrich and Van Neer 1994; Besenval and Desse 1995). While these models are specific to certain regions of South Asia, they have value for similar regions and areas of cultural complexity. The study presented below was part of a much larger research project that attempted to reconstruct archaeological fish processing of the Baluchistan and Indus Valley Traditions of north-western South Asia, *c.* 3300 to 1700 BC (*cf.* Kenoyer 1991).

Previous Research

This research follows earlier pioneering contributions by Stewart and Gifford-Gonzalez (1994) and Stewart (1987; 1991), that attempted to understand fish butchery through skeletal element representation, centred on modern, hunting-gathering-fishing communities in the Lake Turkana region of east Africa. Models developed from Stewart's research have been used in an attempt to understand the use of fish resources by early hominid populations in East Africa. A model for Pakistan offered by Besenval and Desse (1995; Datvian *et al.* 2004; Desse and Desse-Berset 2001) discusses household fish butchery in the Pasni area along the Makran Coast. Major conclusions contend that fresh fish are butchered "in the round," that is, a fish is cut into variously-sized pieces. An additional conclusion is dried

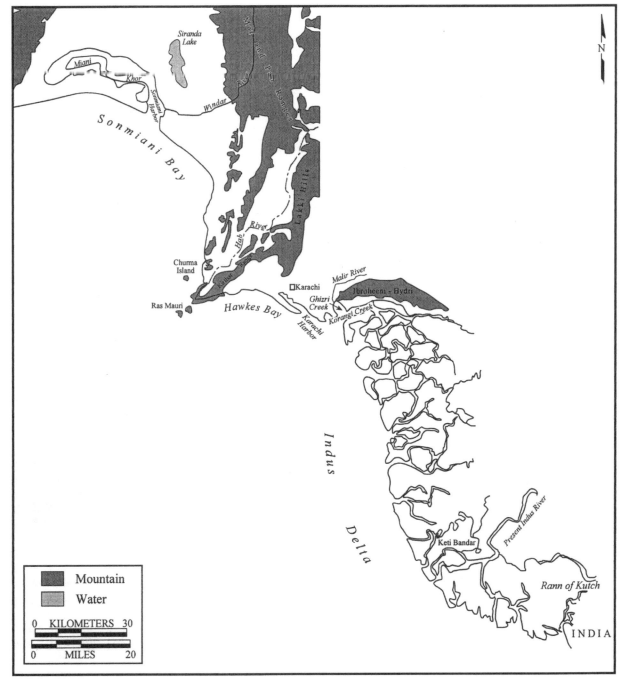

Fig. 11.1 Research locations along coastal Pakistan

fish *cannot* be discerned from fresh fish butchery. However, this study was limited only to a few species observed for a few weeks of observation. The current paper attempts to develop a testable model based on a variety of systemic contexts (fishmongers, urban and village consumers and a variety of market settings) observed over several months between 1992 and 1994.

The Ethnographic Setting of Marine Zone Butchery Systems

Butchery research along the Pakistani marine coast was focused on the village of Abdur Rahman Goth (village) and the fish markets of Karachi (Figs. 11.1 and 11.2), including the West Wharf Fisheries, Lee (or Lea) Market, Empress Market and Kharadar. Abdur Rahman Goth is a small Baluchi-speaking fishing village with a local ice-house, as well as a boat-building operation that constructs commercial as well as traditional vessels (Belcher 1998; Siddiqi 1956). In this village, fish are procured, consumed, and traded. West Wharf is a large, commercial area where the Pakistani fishing fleet lands its catch. Here, the fish are butchered and processed for fresh sale and for freezing. These products are then sold in regional and global markets. A small area of the market processes fish for sale to both individuals and agents buying for hotels and small

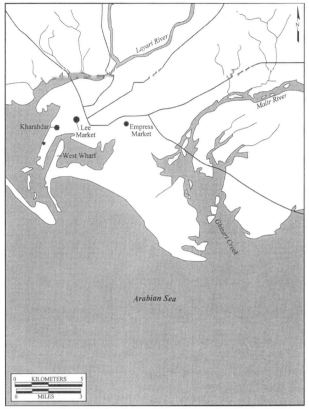

Fig. 11.2 Market locations in Karachi, Pakistan

These observations were either recorded in the form of notes or still photography. During participation, notes were made after events had taken place, while purely observational notations were made during the actual time of the events.

The participant-observation activities were used to develop a series of idealized models that represent a specific faunal assemblage that would represent a specific aspect of human behaviour. It must be emphasized that there are idealized, descriptive models based on specific actions observed at specific points on the pathway from a processing system into the archaeological record. These observational models offer insight into what can be inspected in the archaeological record. However, it is recognized that the archaeological record is a palimpsest of mixed refuse deposits in varying contexts and may not represent specific human behaviours (*e.g.* Binford 1980; Schiffer 1995). Thus, specific behavioural signatures may be masked and obscured by post-depositional human and natural, taphonomic processes. The archaeological record is an epiphenomenon in which several processes, either singly or linked, may produce similar material signatures.

A Model for Provisioning

Provisioning includes the entire range of activities and pathways that allow food to be distributed from its source to the consumers in a human settlement (Meadow 1991; Zeder 1991). In order to understand provisioning properly, it is necessary to examine it as a series of stages defined by categories of labour. In a commercial fishing venture, such as these in Pakistan, several different types of labour are recognized, including: (1) procurement; (2) preparation or processing; (3) transport or distribution; and, (4) cooking. These activities may or may not occur in a linear fashion. Procurement labour includes all activities necessary to extract fish from their natural environment. Preparation or processing labour is defined as any activities associated with butchery and can occur in household or market contexts. Transport or distribution labour involves those efforts of moving materials from a point of procurement to the consumer (usually via central place markets or households). Cooking labour represents domestic activity and involves preparation of meals and eventual discard of fish remains into an initial archaeological context.

In the Buleji/Hawkes Bay area, most of procurement and processing labour is done by village men, while women prepare and serve the meals; women also are responsible for clean-up and discard of meal waste. Village household butchery usually takes place in the home and is done by the women of a household. Most commercial butchery occurs in central place markets, as well as in villages, and is done by men who are full-time fishmongers.

For provisioning, Zeder (1991, 36–37) proposed that the mode of meat distribution, direct or indirect, has a profound effect on the product diversity received by consumers. Direct distribution entails primary access to fish directly from the fisher folk. The content of the catch

restaurants in Karachi. Additionally, small food stands off the market prepare the fish in a variety of manners for immediate, local consumption. Empress Market is one of the 'newer' food markets, built in the 19th century by the British as a separate market for colonial, expatriate residents. This market is still considered an 'elite' market where upper-class Pakistanis and foreign business and consular personnel purchase foodstuffs and other goods. Prices for most goods tend to be more expensive here than in other areas of the city. Kharadar is one of the older markets in Karachi. Here the relatively lower economic and social groups, as well as ethnic Baluch peoples, purchase food, particularly fish.

Ethnoarchaeological Field Work in Pakistan: Methodological Considerations

The purpose of this ethnoarchaeological fieldwork was to record processing and butchery activities at specific locales related to fisher folk, fishmongers, and consumers. Initial research was conducted between January and April 1992 and between December 1992 and April 1993 (Belcher 1994). These preliminary studies were followed by more substantial data collection between October 1993 and August 1994 (Belcher 1998). Using a combined method of participating in various activities of procurement labour and observing activities such as capture, processing and marketing, an understanding of fishing strategies, butchery practices, cooking, and discard was developed.

can vary daily, although more significantly it varies on a seasonal level (Belcher 1998), at least in an ethnographic setting observed for a relatively short period of time. For example, on winter days, a fisher may bring in only one or two fish of a single species, while on a fall, or spring day, several species may be caught. Availability and abundance of particular fish varies throughout the year (seasonally) and year to year (annually), creating a major source of diversity in distribution and consumption (Nikolsky 1963). Seasonal and annual variation is based on fish behaviour, which is in turn reflected in human procurement strategies (Cannon 1996). Additionally, climatic variation and the effects of human predation play a role in changing fish populations in a larger time frame (Fagan 2006). Once caught, decisions related to direct distribution of the fish are made by the fisher folk based on the content of the catch as well as their own food needs and the food needs of their dependents.

For indirect distribution, transport decisions are made by a distributor in order to maximize the efficiency of the distribution system (Zeder 1991, 37). Important distribution factors include distance to consumers, weather, consumer preferences, rate of spoilage, and possible storage techniques; thus, these factors tend to regulate the types of fish present within an indirect mode of distribution. In modern Pakistani fishing villages and boat landing areas, these decisions involve aspects of policy and market preferences rather than personal relationships between individuals (Belcher 1998). Zeder (1991) suggests indirect distribution systems should exhibit less potential product diversity than direct distribution. Three analytical units monitor product diversity and, thus, can be related to distribution systems: (1) species representation; (2) fish size; and, (3) butchery patterns. All of these characteristics are measurable in the archaeological record and provide an operational methodology for interpreting the zooarchaeological patterns visible in the archaeological record.

Butchery occurs in several settings; however, two main areas are of concern here: village household butchery and commercial-based butchery. Village-based butchery usually focuses on households that have more direct contact with the producers/fishers, usually the households of fishers. Commercial-based butchery for fresh fish usually occurs in a central place location, such as a market. Processing for dried fish usually occurs in a village or extra-market location, while dried fish butchery usually occurs in a central place market. The focus of this portion of the study is to understand various aspects of butchery and relate them to possible ideal material correlates that would be found in the archaeological record using the analytical units discussed above.

Through the examination of butchery mark distribution, as well as the species representation and size, it is possible to discern if domestic debris represents that of direct producers/harvesters or indirect consumers. Even if these domestic areas are cleaned regularly, bones of small fishes should remain in some quantity. Based on the ethnographic observations, the position and location of cut marks should be determined by the type of processing that is conducted and the specific orientation of the bones in reference to the direction of cutting force. Butchery marks are usually in the form of chop marks, radiating fractures and crushing of the skeletal elements, called "sharp dynamic point-loading" by Lyman (1987).

Village Household Butchery

Village household butchery primarily occurs with direct home consumption by the producers, their dependents, or other close kin. Specific fish or fish portions tend to be used for home consumption, as other species are highly marketable and would tend to reduce the producer's income. Near-shore fisheries produce a variety of species of fish with only the smaller species or individuals consumed by the fisher folk. Fixed, inshore nets provide large quantities of mugilids and clupeids (mullets and herrings) and are the major source of fish for village families (Belcher 1999). Some of the larger fish are butchered and the body sent to the regional market, while the head will be consumed in the fishing village. Figure 11.3 displays a flow chart that summarizes the decision-making process that occurs during village-household butchery.

Small fish are usually gutted, cooked whole, and then consumed. The heads are chewed and swallowed while the vertebrae are removed during chewing, and discarded. Based on experimental observations from Jones (1986), as well as personal observations of the author, little is left of the cranial remains of these clupeids and mugilids after they pass through the human digestive tract. Whole, medium-sized fish species usually are butchered by gutting and removing the gills. These fish are then split into two halves along the vertebral line, and then chopped into pieces approximately 5cm^2. Sometimes the cut is made along the dorsal surface of the back, and other times the fish is split along the ventral surface from the anus to the head. Heads from large fish are chopped in half along the medial line and then further chopped into smaller pieces to be fried or cooked in a type of stew (called *sālan* or curry).

In summarizing these activities, several ideal assemblages can be discerned. Village household butchery should be dominated by mugilid and clupeid remains. Due to consumption of whole fish by the villagers, these smaller fish are predominantly represented by postcranial remains. Medium-sized fish remains will be present, although not in abundance. Large fish should be represented by a disproportionate amount of cranial remains. Fish bones discarded during and after the meal would eventually be dumped areas outside the main courtyards of the houses. Usually all trash is thrown to either side of the doorway. Small remains that cannot be swept away may be incorporated into courtyard areas or interior house floors, although the majority of other bones are deposited outside in street sediments. Butchery marks will occur on almost all of the cranial elements of the medium and large fish remains, and infrequently on the post-cranial elements. No

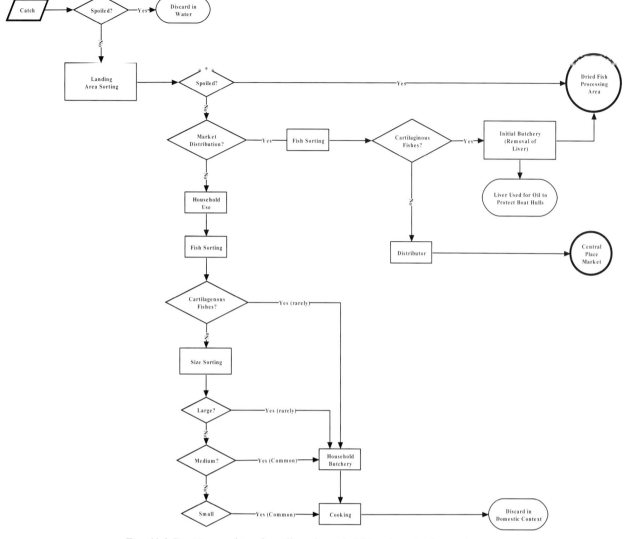

Fig. 11.3 Decision making for village household butchery in the marine zone

cut marks should be present on small fish remains due to the general lack of butchery (except during gutting).

Marine Commercial Butchery

During the 1993–1994 field work in Buleji/Hawkes Bay, a single commercial fishmonger worked in the village area. Thus, when fish are brought ashore, substantial haggling occurred between the fishers and this fishmonger. The fishers did not usually transport their fish the some 20km to the Karachi fish markets as most of them did not own any form of transport. Instead, the fishmonger bought the fish and transported them to the markets of West Wharf and Kharadar, where they are auctioned to various buyers or sold to other fishmongers in various markets, such as Lee or Empress. During the lean winter months (in general, few fish were caught), the fishmonger would often loan money to fishermen to buy necessities such as diesel, new nets, or pay for repairs on their boats and engines. The fishmonger would also loan the fishers money for weddings, funerals, or holiday celebrations. Through this method, the fishermen became obligated to sell to the fishmonger during the summer months, when fishing is more productive.

The fishmonger selects only certain fish based on species, colour, and appearance. He did not want to purchase spoiled fish, although he would give a few rupees to a fisher for an entire catch of spoiled fish. Due to relatively high water temperature, particularly in the summer, fish would spoil and rot rather quickly. This spoiled fish was often sold to a local *malik* (owner, entrepreneur), who controls the village dried fish operations.

Two products are the focus of commercial-based butchery: dried fish and fresh fish. Dried fish are processed and prepared in villages and central place markets. Usually dried fish is traded from these coastal villages to inland villages and to the regional centres, such as Ormara, Pasni, Gwadar, and eventually Karachi, where it is disseminated to the interior region, such as mountainous Baluchistan and interior Sindh (Hyderabad, Sukkur). Based on the ethnographic observations summarized below, Figures 11.4 and 11.5 display a flow chart that summarizes the decision-making process that occurs during commercially-based fish

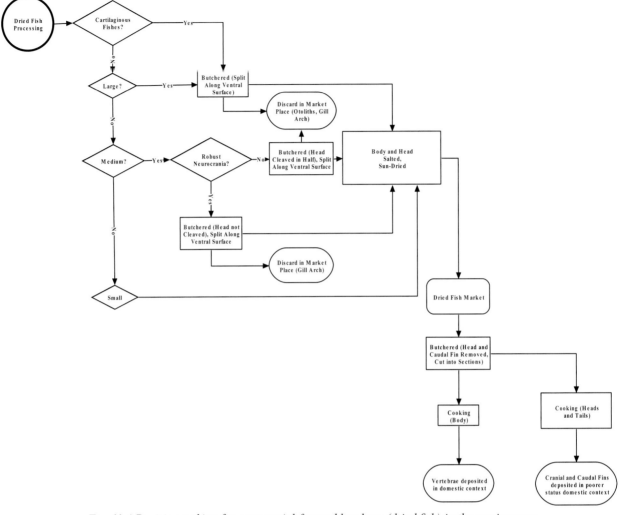

Fig. 11.4 Decision making for commercial-focused butchery (dried fish) in the marine zone

butchery, including dried fish processing/butchery, as well as fresh fish butchery.

Dried Fish Processing

Usually only particular species are dried and salted for consumption. Small fish (less than 25cm long) such as mullets, herrings, and small drums are dried directly in the sun; these small fish are often used as poultry food and rarely as human food. Other, larger fish are prepared for drying initially by gutting, with the gills and entrails discarded, and the body and head are split lengthwise along the belly. In this manner, the fish can be laid flat (butterflied) and stacked for ease in drying and salting. Often a small axe or large cleaver/knife is used in this operation (Fig. 11.6). Usually the vertebral column lies along the left flank of the body and the neurocranium will be split lengthwise to lie flat. Slits are cut along the inside portion of the flanks to allow salt to be rubbed into the interior flesh in a form of dry curing, using salt and sun. However, if the head possesses a fused neurocranium, such as marine catfish (Ariidae), grunters (Haemulidae), or drums (Sciaenidae), the head is not cleaved in half, instead only the body is butterflied and split.

Remains from dried fish processing should possess distinctive sets of cut marks that distinguish them from other processing strategies. Due to the manner in which the fish are processed, cut marks should be present along the medial sides of ribs and cranial elements of particular species, and such species should exhibit chop marks and trauma along the medial surface of the bones. Additionally, during dried fish preparation, very few skeletal elements would be intentionally deposited in the archaeological context of the processing site as the fish dries, but certain elements will disarticulate from other skeletal elements and separate from the dried flesh. These elements can become incorporated into the archaeological record, although the dried fish body will be taken to another locale and sold or consumed. Generally, these elements include those of the gill arch/hyoid arch (branchiostegal region) and otoliths from species possessing non-fused neurocrania.

Fresh Fish and Dried Fish Butchery

For fresh fish, taxonomic class is an important variable in butchery. Different systems and patterns of butchery

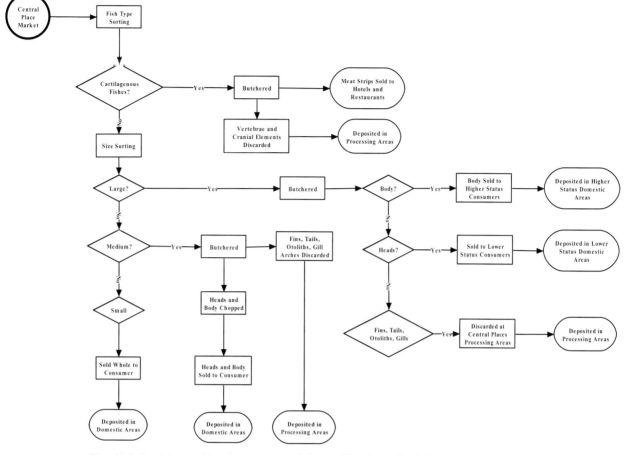

Fig. 11.5. Decision making for commercial-focused butchery (fresh fish) in the marine zone

are used if a fish is a cartilaginous fish (Chondrichthyes), such as a shark or skate/ray, or a bony fish (Osteichthyes). Due to general religious restrictions, cartilaginous fish are not generally sold in the fish markets of Karachi, but are usually destined for sale to Southeast Asia and Sri Lanka. However, that said, the author has seen numerous examples of local fisher folk consuming shark meat. For sharks and rays, the heads, entrails and vertebral column are discarded at the market. The tails and fins are prepared for sale to foreign markets, while the meat is often cut into strips and sold as fried fish in the bazaars of Karachi and to hotels frequented by foreigners.

Bony fish are butchered in a various styles dependent on size. Based on both interviews and observation, three sizes of fresh fish are noted: small (<25cm); medium (between 25cm and 1m), and large (>1m). Small fish usually are sold whole to the consumer (Fig. 11.7). For both medium and large fish, the gills are removed along with the entrails, tails and fins (Fig. 11.8). For larger fish, the head is severed from the body and the two sections usually are sold separately. This follows the pattern of butchery "in the round" as suggested by Besenval and Desse (1995).

With medium-sized fish, the head and body are cut into pieces; gills and fins are discarded at the butchery site. Thus, medium fresh fish are butchered into small pieces used for cooking, segmenting the skeleton into similar-sized pieces. Cut marks and chop marks will primarily occur on the cranial elements with few cut marks on the vertebral column, except to separate the body into several portions. Fin rays and spines, the last few caudal vertebrae, pharyngeal plates and teeth (removed with the gills), as well as the distal and midshaft portion of catfish (Siluriformes) pectoral spines, characterize the bone assemblages of fresh fish processing areas.

'Stylistic' differences in butchery also appear to be associated with various ethnic groups and economic classes. Therefore, these socio-economic variables should affect the distribution of discarded materials throughout Karachi and the outlying villages. For example, particular ethnic groups in Karachi, such as highland Baluch and interior Sindhis, consume some quantity of dried and salted fish. Thus, these ethnic households should contain assemblages reflecting dried fish processing. Lower economic groups tend to buy less desirable cuts of fish, which include heads of large fresh fish, as well as heads and tails of dried fish. These cuts tend to possess a large quantity of bone and little, actual flesh. Middle income groups normally buy medium-sized fish (heads and bodies) as well as large fish bodies. Thus, middle income groups select cuts that have a larger amount of flesh present. Upper economic groups tend to frequent specific markets, such as Empress Market. Particular species, such as white pomfret (*Pampus argenteus*) and the Indian shad

Fig. 11.6 Preparing a fish carcass for drying at Abdur Rheman Goth (Village), Buleji, west of Karachi. Photograph by the author.

Fig. 11.7 This Baluch gentleman specializes in the butchery of medium-sized fish, especially marine catfish (Ariidae), and grunters (Haemulidae) in Lee Market. Photograph by the author.

Fig. 11.8 This Sindhi fishmonger specializes in the sale of smaller fish in Lee Market. Photograph by the author.

(*Tenualosa ilisha*) are considered highly desirable and can only be purchased by upper income families. In these markets, fish are frequently filleted and only the flesh is taken to the domestic areas. Often filleted skeletons are sold to the lower economic groups.

Ideally, various domestic areas could be defined by bone assemblages. Specifically, heads and tails would appear more commonly in lower socio-economic areas of an urban centre, while vertebrae and heads are more likely to occur in middle income domestic contexts. Certain species, due to their desirability, probably will be deposited only in higher status domestic areas; nevertheless, bone assemblages would appear similar to the middle economic areas.

Fish filleting was only observed in higher status fish markets, especially those that catered to foreign populations. While filleting can be seen in the markets, it does not appear to be a common manner in which fish are butchered in a South Asian context. However, as noted above for fillets, the bone materials evidence may be deposited at the processing areas or in lower income contexts as these items are given or taken away by lower income families to create soups. Processing areas are usually located in market areas of a regional centre, away from domestic quarters. Often these filleted materials would be butchered into small packets of bone and meat. Thus, it may be difficult to distinguish the consumption of fillets from whole fish butchery as the skeletal element and butchery mark signatures would be similar.

Particular skeletal elements may reveal signatures related to relationships between consumers and producers. Based on observations of modern Pakistani fisher folk and interviews, the sharp fin spines of catfish are broken by the fisher folk to avoid injury; however, for bagrid (Bagridae), ariid (Ariidae), and some silurid (Siluridae) catfishes, the spine's external membrane can hold these broken pieces together prior to initial transport. At fish markets, these spines are removed near the proximal articular head (see Fig. 11.3). Thus, deposits reflecting a direct mode of acquisition from the producer (fisher folk) may contain the pointed, distal fragments, the proximal articular head with a good portion of the mid-area, as well

Butchery Style	Systemic Context (Origin)	Cut Mark Location	Element Representation in Archaeological Context	Ideal Archaeological Context (Depositional Context)
Village Household Butchery				
Fresh Fish				
Small	Domestic	None	Postcranial dominates	Domestic
Medium	Domestic	Cranial - Lateral	Cranial and Postcranial	Domestic
Large	Domestic	Cranial - Lateral	Cranial dominates	Domestic
Commercial-Based Butchery				
Fresh Fish				
Small	Central Place (Market)	None	Not present	Processing Area
		None	Postcranial dominates	Domestic Area
Medium	Central Place (Market)	Lateral on gill arch/hyoid arch region	Branchial Arch, Fin Rays, Lower Hyoid Arch	Processing Area
		Lateral on Cranial and Post Cranial	Cranial and Postcranial	Domestic Area
Large	Central Place (Market)	Lateral on gill arch/hyoid arch region	Branchial Arch, Fin Rays, Lower Hyoid Arch	Processing Area
		Lateral on Cranial and Postcranial	Cranial and Postcranial	Domestic Area 1 (Higher Socioeconomic Class)
		Lateral on Cranial and Postcranial	Predominantly Cranial	Domestic Area 2 (Lower Socioeconomic Class)
Dried Fish Processing				
Robust Crania	Central Place and Village	Medial along ventral ribs and vertebrae	Cranial/Postcranial	Domestic Area
Non-Robust Crania	Central Place and Village	Medial along ventral ribs, vertebrae and cranial areas	Cranial/Postcranial	Domestic Area
Dried Fish Butchery				
Small	Central Place	None	Postcranial	Domestic Area
Medium to Large	Central Place	Lateral	Cranial/Postcranial	Domestic Area 1
'Scraps'	Central Place	Lateral	Caudal Vertebrae/Cranial	Domestic Area 2

Table 11.1 Butchery style and element representation

as complete or virtually complete pectoral and dorsal fin spines. Acquisition of fish through indirect methods, such as a market place, would be reflected by deposits containing primarily articular head fragments from the pectoral or dorsal spine.

Summary

The ethnographic observations in various idealized systemic contexts are outlined in Table 11.1. This table summarizes the ethnographic analyses grouped by systemic and archaeological contexts. The major categories divide the activities into archaeologically measurable units of analysis, such as cut mark distribution, skeletal element representation, and fish size. However, it must be emphasized that these are idealized contexts. The archaeological record is rarely so exact and a variety of taphonomic and preservational processes affect the current archaeological record. However, these idealized models offer a valuable system for understanding fish processing and distribution in a South Asian context.

Village and Commercial Butchery

Village household butchery is highly variable, not only in terms of skeletal element representation, but also in terms of cut-mark distribution. Cut-mark distribution should be highly variable, probably due to the effort of butchering the fish into small, bite-size pieces, whereas in commercial butchery, cut-mark distribution appears more standardized. Village household butchery tends be fairly distinctive in terms of element representation. Direct and indirect modes of distribution are quite dissimilar in terms of skeletal representation, as well as cut mark location.

Urban households that obtain fish from central place markets are problematic. Fish butchered at a central place market tend to reflect standardized butchery. However, if the whole fish are taken to the home and butchered in a domestic setting, these assemblages should reflect a more random or individualized butchery pattern. Small fresh fish tend to be purchased whole at markets and will be butchered at home. Thus, in reference to smaller fish, it is difficult, using only size, to determine whether or not fish were obtained through a direct or indirect mode of distribution.

Dried fish butchery tends to produce cut marks on cranial and caudal vertebrae. Of course, also present on these remains are those cut marks that reflect dried fish processing tasks. This suggests that even though dried fish butchery activities are similar to fresh fish butchery, dried fish butchery may be distinguished from fresh fish butchery by cut marks derived from the processing strategies used to dry and salt the fish.

Dried Fish Processing

Dried fish processing differs significantly from other forms of processing and butchery in terms of cut mark location. Cut marks are principally located on the medial side of the bone – that is, along the inside surfaces. This occurs due to the action of splitting the fish for salting and/or drying in the sun. Although, other forms of butchery create medial cut marks, dried fish processing produces a greater frequency of these medial cut marks.

Discard Patterns

An important factor in modern butchery is scale. The largest scale is seen in commercial butchery at West Wharf in Karachi, where heads are discarded, whereas at the 'non-elite' markets and in village household butchery, discard is characterized by caudal vertebrae (tail fins), fin spines, various dorsal structures, as well as the gill arch. Also at these local markets, heads are butchered and sold to lower income groups to be eaten rather than discarded. This large scale of butchery is more common at West Wharf than in the other markets around Karachi because West Wharf is the primary commercial fishing and trade port for Pakistan.

Heads are usually detached at the base of the skull, or basioccipital region, often removing the rear portion of the neurocranium from the remainder of the head. Thus, the rear portion of the head is attached to the body. Heads often are removed by separating the head from beneath the gill plates, thus leaving the pectoral girdle attached to the body portion of the fish. Local fishmongers state that the pectoral girdle helps keep the body from getting crushed during shipment to a distant market. This is especially true for catfish (Siluriformes, such as Siluridae, Bagridae, and Ariidae). This pattern probably represents activities that were not present in the past, as this pattern requires the use of ice as a preservative.

After a meal has been prepared, often all the fish are not consumed and will be kept for later meals. Thus, not all the remains from a single cooking activity will be discarded at the same time. This results in one to several discard events that may even occur in separate dumps in a domestic area. This pattern suggests that, in a domestic setting, skeletal elements may be processed and eventually discarded over a period of several days, forming a palimpsest deposit that does not reflect individual meals.

Conclusions

This paper presents a preliminary model based on ethnographic observations for various contexts for village-level and urban-centred butchery and processing activities. Due to initial comments by an anonymous reviewer, the author reviewed field notebooks and still photographs to define these idealized model; thus, some of the comments have been changed from those presented in Belcher (1998). This research has attempted to build on the work of previous researchers who focused on hunter-gatherer-fishers as well as pastoral groups. These idealized models are based primarily on observations over several months in several different behavioural contexts. From these contexts, the types of archaeologically-available information in terms of species representation, skeletal element representation, and cut mark distribution are predicted.

The next stage of research is to continue the analysis and collection of skeletal element assemblages from the various systemic contexts. This will allow these idealized situations to be examined in an archaeofaunal system. This was done with limited success in Belcher (1998). Furthermore, excavations must be done in the various contexts in small villages, processing areas, and urban and rural butchery locales. This will allow a more definite model that reflects preservation and taphonomic processes to be examined in more detail.

Acknowledgements

This research was funded by the Pakistan Education Foundation through a Fulbright Student Grant, a National

Science Foundation Dissertation Improvement Grant (Award No. 9306645), the Harappa Archaeological Research Project and private donors. Facilities for analysis were provided by the Centre for Excellence in Marine Biology, University of Karachi, the Harappa Archaeological Research Project, the French Archaeological Mission in Pakistan as well as the Department of Anthropology, University of Wisconsin-Madison. This research would not have been possible without the cooperation and help of the fisher folk of Buleji/Hawkes Bay and the fishmongers of Empress, Lee, and Kharadar Markets. With humour and some well-understood annoyance, these individuals allowed me to intrude on their lives with observations, chatting, numerous questions and photography. Figures 11.1 and 11.2 were redrafted by Ms. Echo Funk based on original maps by W. Belcher. Much necessary organizational and editorial comments were graciously given by Laura Miller and Linda Kohstaedt. I would also like to thank the editors of this volume, as well as the anonymous reviewer, all of whom greatly enhanced the content of this chapter. The anonymous reviewer forced me to re-examine many of my original observations. I (once again) read through my field notes, and reviewed my still photographs.

References

Belcher, W. R. (1994) Butchery practices and the ethnoarchaeology of South Asian fisher folk. In W. Van Neer (ed.) *Fish exploitation in the past. Proceedings of the 7th Meeting of the ICAZ Fish Remains Working Group.* Annales du Musée Royal de l'Afrique Centrale, Sciences Zoologiques 274, 169–176. Tervuren.

Belcher, W. R. (1998) Fish Exploitation of the Baluchistan and Indus Valley Traditions: An Ethnoarchaeological Approach to the Study of Fish Remains. Ph.D. Dissertation, Department of Anthropology, University of Wisconsin-Madison. Ann Arbor, MI, University Microfilms.

Belcher, W. R. (1999) The ethnoarchaeology of fishing in a Baluch village (Pakistan). In H. P. Ray (ed.). *The Archaeology of Seafaring: The Indian Ocean in the Ancient Period*, 22–50. Delhi, Pragati Publications.

Besenval, R. and J. Desse (1995) Around or lengthwise: fish cutting-up areas on the Baluchi coast (Pakistani Makran). *The Archaeological Review* 4(I & II), 133–149.

Binford, L. R. (1980) Willow smoke and dogs' tails: Hunter-gatherer settlement systems and archaeological site formation. *American Antiquity* 45, 4–20.

Brinkhuizen, D. C. (1994) Some notes on fish remains from the late 16th century merchant vessel Scheurrak SO1. In W. Van Neer (ed.) *Fish exploitation in the past. Proceedings of the 7th Meeting of the ICAZ Fish Remains Working Group.* Annales du Musée Royal de l'Afrique Centrale, Sciences Zoologiques 274, 197–205. Tervuren.

Bunn, H. T. (1994) Early Pleistocene hominid foraging strategies along the ancestral Omo River at Koobi Fora, Kenya. *Journal of Human Evolution* 27, 247–266.

Bunn, H. T. (2001) Hunting, power scavenging, and butchering by Hadza foragers and by Plio-Pleistocene Homo. In C. Stanford and H. T. Bunn (eds.) *Meat-Eating and Human Evolution*, 199–218. Oxford, Oxford University Press.

Butler, V. L. (1993) Natural versus cultural salmonid remains: origin of the Dalles Roadcut bones, Columbia River, Oregon, USA. *Journal of Archaeological Science* 20, 1–24.

Butler, V. L. and R. A. Schroeder (1998) Do digestive processes leave diagnostic traces on fish bones? *Journal of Archaeological Science* 25(10), 957–971.

Cannon, A. (1996) Scales of variability in northwest salmon fishing. *In* M.G. Plew (ed.) *Prehistoric Hunter-Gatherers Fishing Strategies*, 25–40. Boise, Boise State University.

Colley, S. M. (1990) The analysis and interpretation of archaeological fish remains. In M. B. Schiffer (ed.) *Archaeological Method and Theory*, Vol. 2, 207–253. Tucson, University of Arizona Press.

Davtian, G., Desse, Y. and Desse-Berset, N. (2004) Occupation du littoral du Makran (Bélouchistan, Pakistan): données archéologiques, contraintes environnementales, apports de la géomatique. Actes des XXVe Rencontres Internationales d'Archéologie et d'Histoires d'Antibes. *Temps et espaces de l'homme en société, analyses et modèles spatiaux en archéologie*, 431–441. Antibes.

de Jong, T. (1994) Fish consumption at Eindhoven Castle: archaeological remains versus historical sources. In W. Van Neer (ed.) *Fish exploitation in the past. Proceedings of the 7th Meeting of the ICAZ Fish Remains Working Group.* Annales du Musée Royal de l'Afrique Centrale, Sciences Zoologiques 274, 129–137. Tervuren.

Desse, J. and N. Desse-Berset (2001) Ancient exploitation of marine resources on the Makran Coast. Paper presented at the Madison South Asian Conference, University of Wisconsin, Madison.

Fagan, B. (2006) *Fish on Friday*. New York, Basic Books.

Hoffman, R. C. (1994) Remains and verbal evidence of carp (*Cyprinus carpio*) in medieval Europe. In W. Van Neer (ed.) *Fish exploitation in the past. Proceedings of the 7th Meeting of the ICAZ Fish Remains Working Group.* Annales du Musée Royal de l'Afrique Centrale, Sciences Zoologiques 274, 139–150. Tervuren.

Hoffman, B. W., Czederpiltz, J. M. C. and M. A. Partlow (2000) Heads or tails: the zooarchaeology of Aleut salmon storage on Unimak Island, Alaska. *Journal of Archaeological Science* 27, 699–708.

Jones, A. K. G. (1986) Fish bone survival in the digestive systems of the pig, dog and man: some experiments. In D. C. Brinkhuizen and A. T. Clason (eds.) *Fish and Archaeology*. British Archaeological Reports, International Series 294, 53–61. Oxford.

Kenoyer, J. M. (1991) The Harappan Phase of the Indus Valley Tradition of Pakistan and Western India. *Journal of World Prehistory* 5, 331–385.

Locker, A. (2001) *The Role of Stored Fish in England, 900 – 1750 A.D.: the Evidence from Historical and Archaeological Data*. Sofia, Bulgaria, The Publishing Group.

Lubinski, P. M. (1996) Fish heads, fish heads: an experiment on differential bone preservation in a salmonid fish. *Journal of Archaeological Science* 23, 175–181.

Lupo, K. (2001) On the archaeological resolution of body part transport patterns: an ethnoarchaeological example from East African hunter-gatherers. *Journal of Anthropological Archaeology* 20, 361–378.

Lyman, R. L. (1987) Archaeofaunas and butchery studies: a taphonomic perspective. In M. B. Schiffer (ed.), *Advances in Archaeological Method and Theory*, Vol. 10, 249–337. New York, Academic Press.

Meadow, R. H. (1991) Faunal remains and urbanism at Harappa. In R. H. Meadow (ed.) *Harappa Excavations 1986–1990: A*

Multidisciplinary Approach to Third Millennium Urbanism, 89–106. Madison, WI, Prehistory Press.

Nicholson, R. A. (1996) Bone degradation, burial medium and species representation: debunking the myths, an experiment-based approach. *Journal of Archaeological Science* 23, 513–533.

Nikolsky, G. V. (1963) *The Ecology of Fishes*. New York, Academic Press.

Raychaudhuri, B. (1980) *The Moon and Net*. Calcutta, Anthropological Survey of India.

Richter, J. (1986) Experimental study of heat induced morphological changes in fish bone collagen. *Journal of Archaeological Science* 13, 477–481.

Schiffer, M. B. (1995) *Behavioral Archaeology: First Principles*. Salt Lake City, University of Utah Press.

Siddiqi, M. I. (1956) *The Fishermen's Settlements on the Coast of West Pakistan*. Kiel, Geographischen Instituts der Universität Kiel.

Stewart, K. M. (1987) *Fishing Sites of North and East Africa in the Late Pleistocene and Holocene*. BAR International Series 521. Oxford, British Archaeological Reports.

Stewart, K. M. (1991) Modern fish bone assemblages at Lake Turkana, Kenya: a methodology to aid in recognition of hominid fish utilization. *Journal of Archaeological Science* 18, 579–603.

Stewart, K. M. and D. Gifford-Gonzalez (1994) An ethnoarchaeological contribution to identifying hominid fish processing sites. *Journal of Archaeological Science* 21, 237–248.

Wendrich, W. Z. and van Neer, W. (1994) Preliminary notes on fishing gear and fish at the late Roman fort at 'Abu Sha'ar (Egyptian Read Sea coast). In W. Van Neer (ed.) *Fish exploitation in the past. Proceedings of the 7th Meeting of the ICAZ Fish Remains Working Group*. Annales du Musée Royal de l'Afrique Centrale, Sciences Zoologiques 274, 183–189. Tervuren.

Willis, L. M., Eren,, M. I. and Rick, T. C. (2007). *Experiments in Fish Butchering: Implications for Bone Modification and Taphonomy*. Paper presented at the 72nd Annual Meeting of the Society for American Archaeology. Austin, Texas.

Zeder, M. A. (1991) *Feeding Cities*. Washington, Smithsonian Institute Press.

Zohar, I. and Dayan, T. (2001) Fish processing during the Early Holocene: a taphonomic case study from coastal Israel. *Journal of Archaeological Science* 28, 1041–1053.

Food Preparation and Consumption

12. Ethnozooarchaeology of butchering practices in the Mahas Region, Sudan

Elizabeth R. Arnold and Diane Lyons

This paper presents an ethnozooarchaeological investigation of livestock production, management, and slaughter practices in the town of Delgo in the Mahas Region in the Northern Province of Sudan. The research objective was to determine the relationship between men's butchery and women's cookery practices. Data is presented from butchering events for two household sponsored feasts. This data was used to generate taphonomic expectations that were then compared to a sample of discarded bone recovered from five village middens. The study further investigates household production objectives for livestock in relation to household structure. Interviews focused on each household's membership composition, and the number and type of domestic animals kept by each household. Observation of butchering and feasting events was combined with a consideration of taphonomic processes that result from the observed practices. The study concludes that women have a strong influence on men's butchering practices. Because women are responsible for all cooking and presentation of food, their interests dictate the size of butchered bone elements that are processed. This type of data lends itself to examinations of feasting events that are linked to socioeconomic competition and status and to the roles that women play in feasting activities. For instance, whose status is negotiated at feasts? Males identified as leaders and hosts of the feast, or the females who have not only prepared the food but also dictated its preparation from the living beast which they have reared for slaughter?

Keywords: ethnozooarchaeology, butchering, slaughter patterns, sheep/goat, gender, feasting

Introduction

Ethnozooarchaeology is the study of contemporary material practices of human–animal interaction in ways that are useful in assisting archaeologists in interpreting the past. Presented is one aspect of the Sudan Doka Project that was conducted between September and October 2005. This project was funded by grants from the University of Calgary (Dr. Diane Lyons) and Wenner Gren (Dr. A. C. D'Andrea, permit holder). Arnold initiated the ethnozooarchaeological investigation of livestock production, management and slaughter practices in the town of Delgo with the assistance of Lyons. Delgo is a town of approximately 2500 people in the Mahas Region in the Northern Province of Sudan. The town is located on the east bank of the Nile River between the second and third cataracts (Fig. 12.1). Beyond the fertile influence of the Nile, the surrounding environment is desert. All Mahas households investigated were Muslim, and household economies in Delgo were based in a combination of farming and the income of family members who were employed in local government offices, small businesses or in other countries, particularly Saudi Arabia.

The initial goal of the study was to determine the

Fig. 12.1 Map of study area

interrelationship between the social and functional contexts of animal husbandry, butchering practices and food preparation. This is an area of investigation that several researchers (David and Kramer 2001; Gifford-Gonzalez 1993) argue has received far too little attention. This study is a preliminary investigation of the interaction between women's culinary and men's butchering practices in a rural town setting. Over the course of the investigation, it was possible to observe and to record two feasting events, each of which included the slaughter of a sheep. While there is still little agreement within the anthropological literature on a standard definition of feasting, it is defined here (after Dietler and Hayden 2001, 3) as an event that is constituted by the communal consumption of food and/or beverages in contexts of interaction that are differentiated from everyday meals. An important element of feasts is the consumption of meat dishes. Although women are generally perceived as contributing to feasts as cooks, this study suggests that women play a greater role in decision making in food production than just cookery. Based on interviews and observation of Mahas households, it was found that while only men are responsible for the act of butchering and only women prepare food in domestic contexts of consumption, it is women's cooking practices that largely dictate the procedures for butchering.

A faunal assemblage was collected randomly from five communal domestic middens in the town of Delgo. This assemblage made it possible to further explore butchering patterns as well as patterns of discard and other taphonomic processes in the community. Based on this small sample, it appears that feasting practices are evident in the middens investigated.

Background and Research Questions

Women's role in livestock keeping is not confined to cookery. Recent studies by the Food and Agriculture Organization of the United Nations (FAO) have shown that rural women, both as paid and unpaid labour, play a significant role in agricultural and food production and in household food security (FAO 1995; Bravo-Baumann 2000). In Sudan, women provide one-third of the labour required to sustain agricultural production. Although men head the majority of rural households, the number of women-headed households is increasing as a result of male migration, widowhood and divorce. The percentage of female-headed households recorded in Sudan is 23.8% (second only to Pakistan). Women spend long hours every day in crop and livestock production and often are responsible for all aspects of animal husbandry, with the exception of herding and marketing. These responsibilities include feeding and watering the animals, gathering fodder, providing care for small ruminants, rabbits and poultry, cleaning of stables, collecting dung for fertilizers and fuel, and caring for the sick, pregnant and lactating animals (including milking, making butter and cheese, and breeding animals). With women's extensive involvement in the care and production of animals, it is expected that women's activities will be evident in ways that are observable in the archaeological record (FAO 1995; Bravo-Baumann 2000).

Gender differences in food production and preparation are recognized in previous ethnoarchaeological research amongst hunter gatherer groups. In these studies many researchers note that butchering decisions are often dictated by how meat and other products from the kill are intended to be utilized and processed (Binford 1978; Bunn *et al.* 1988; Gifford-Gonzalez 1991; 1993; O'Connell *et al.* 1988; Yellen 1977). Studies highlight that while men perform the slaughtering, women do the cooking and presentation of the final cuisine. As a result, slaughter and butchering patterns, and preference in cuts of meat may be dictated by decisions made by women that are constrained by such elements as vessel size, cooking fuel and the availability and preparation of plant foodstuffs (Gifford-Gonzalez 1993). It is anticipated that similar processes are present in other subsistence economies where men's butchering and women's cooking are integrated processes.

The production objectives for livestock herds can be variable depending on household structure. A family with young children may prioritize and maximize milk production while families with older children may emphasize marketable animals (Niamir 1982). Priorities may include increasing herd size, increasing milk yield, maintaining an appropriate herd structure for short and long term reproductive success, and ensuring disease resistance through selective breeding (Monod 1975). While the priority given each goal will vary depending on a household's structure and other factors, Dahl and Hjort (1976) emphasise that herds will always be managed for growth regardless of other economic goals.

Methodology

The documentation of basic household composition and livestock data was recorded through unstructured interviews with the aid of an interpreter. Informants were asked questions regarding livestock ecology and management procedures including herd composition and structure, feeding strategy and its seasonal aspects, level of mobility, labour requirements, and stock uses. General information was also obtained regarding household structure and the number, age and sex of animals owned and utilized by the household; foddering practices and medicinal care of animals; and who was responsible for animal care. Preparation of food products produced from the animals and the gendered division of labour of butchering and food preparation was also examined through both interview and observation.

Two slaughtering episodes were directly observed. The first was the slaughter of a sheep in a domestic setting for the purposes of a large family gathering. The second was in a commercial setting that involved the slaughter of a cow by the local butcher for market day. Commercial butchering consists of slaughtering two cows (obtained locally) per week on Monday and Thursday. Analysis of the commercial butchering is currently underway by one of the authors (Arnold). Each event was extensively photographed and documented. It was possible to observe and record the subsequent cooking and presentation of the butchered sheep to the guests. A second feast was observed and documented but on this occasion it was not possible to observe the slaughter of the second sheep.

A collection of animal bones from five communal middens in different parts of Delgo was analyzed in order to compare the small sample of observed practices with a larger sample of butchered bones. Faunal material was analyzed for species, element, sex (if possible), age and taphonomy. Photographs were taken of all butchered faunal remains.

Results

The dominant livestock resources in the Sudan include cattle, sheep, goats and camels. Delgo households did not keep camels although donkeys were present as beasts of burden. The absence of camels was explained by informants as an ethnic characteristic, as the Mahas do not keep camels. Cattle are predominantly kept for milk or meat and these preferences are locally determined. Sheep are the Sudan Desert type and goats are mostly the large, black Nubian type. Both are kept for milk in Delgo.

Household Structure and Production

Interview questions focused on human household composition and the number and type of domestic animals kept by the household. Other questions focused on products obtained from the animals, such as meat, cheese and yoghurt, and the details of obtaining these products, including how often animals were milked. Table 12.1 presents a summary of the number of animals kept by the households. Ten households provided information for this investigation constituting a small sub-sample of the population.

Seven of the ten households kept both sheep and goats, and three households only had one of the two species. Sheep predominate in the overall sample, comprising 48% of the large stock (excluding chickens, pigeons and rabbits). Goats make up 38% of the sample and cattle 14%. The average number of small stock per household is between four and five animals. In addition, six families kept one to four cattle.

The main product from livestock is milk; meat is secondary. There is considerable variation in milk production and utilization in households. This trend is not directly linked with household composition, and there was no discernible pattern for keeping milk-producing animals only for the nutritional benefits that these products provide to children. For example, in one household with one small child, goats were milked both in the morning and in the evening. However, this pattern was also noted in households without children. In other households with no children, goats were milked only in the evening. One element of consistency was that everyone in the households drank milk. One informant reported that in particular, the older women of that household drank milk (there were no young children present). Families that own cows will preferentially use cow's milk, milking both morning and evening, regardless of the number of children in the household.

Cheese and yoghurt are common products made from the milk of both goats and cows, although it was not produced in all households. In addition, it was noted that with children present in the household, less milk was used for these secondary products, with more milk reserved for consumption by the children. The frequency of yoghurt making was approximately once a week. This finding contrasts with that of Niamer (1982) such that the presence of children *per se* does not affect milk production, but the presence of children does affect women's considerations in how the milk is processed.

When livestock have calves or kids, allowances are made to provide sufficient milk to the offspring. In one situation, with the only cow in the household with calf, milking in the morning was for household consumption, while the calf was allowed to feed in the evening. In a second instance, a wealthy man with four wives had three cattle including two cows. Only the younger cow was milked for the households. This occurs in the evening and the milk is shared amongst the wives. At the time of the study, the older cow was pregnant, and informants stated that all milk would be reserved for the calf. This consideration is evident

Household	Number of Cattle	Number of sheep	Number of goats	Number of chickens	Number of rabbits	Number of pigeons	Number of camels
1	0	0	4	8	2	0	0
2	0	4	2	7	0	0	0
3	3	7	8	12	0	present	0
4	4	13	6	0	0	0	0
5	2	10	6	present	0	present	0
6	0	4	4	0	0	0	0
7	3	0	2	0	0	0	3
8	3	8	0	0	0	0	0
9	0	4	5	0	0	0	0
10	1	5	6	0	0	0	0

Table 12.1 Summary of livestock per household

in a third household with three cows, consisting of two pregnant cows and a young calf. These animals were not milked at all and provided no milk for the household. In a fourth household, with sheep/goats and cattle, the cow was pregnant and milk was left for the calf. With the exception of the first two examples, none of these households had small children. It was noted in one instance that the goats do not give much milk and so powdered milk was used as a substitute. This is consistent with observations presented by Dahl and Hjort (1976) that the primary management goal is to maximize growth of the stock.

Sheep and goat are most often penned together and are brought fodder by the women of the household. It was noted that stock were let out to graze every two to three days. Cattle are not allowed to graze and are tethered and fed with fodder in a separate area away from sheep and goat. Sorghum and maize stalks, berseem (*Trifolium alexandrinum* L.) and date palm leaves (*Phoenix dactylifera*) are all used as fodder.

It was reported that sheep are the preferential animal for slaughter for feasts. There is some variability in the prime age of animals at time of slaughter; one household reported a prime age of 6–7 months for slaughter while most others report a prime age of 1 year. Male animals are preferentially selected for butchering to conserve females for reproduction and milk production. Slaughter took place on special occasions such as the end of Ramadan, or when special guests were visiting. The husband is responsible for butchering, and it was noted that the wife did not engage in any cutting of the carcass. Today, as many men work abroad, there are a greater number of households without men for portions of the year. In these cases a male neighbour or relative performs the butchering for such households. Butchering occurs near the context of consumption. In Delgo commercial butchering is near the market, and the sheep slaughter observed during this study was performed outside of the owner's house just a few meters from the back entry to the women's area and kitchens, and on the east side of the house, away from the main road and pathways.

It was also observed that several families kept chickens and pigeons. Chickens were kept mainly for eggs but were sometimes eaten. Pigeons are kept for meat, and several households had several dozen birds that were eaten only occasionally. Only one household kept rabbits for meat.

Butchering

It was possible to observe domestic butchering of a male sheep. Butchering occurred outside of the house at a side doorway and away from pathways or roads. The front feet were bound and the animal's throat was cut while the animal faced east as the sun rose. A sack was placed under the neck to catch some of the blood. The sheep was left for several minutes to bleed out and to stop moving before butchering began. First, the metatarsal of one leg was skinned. A cut was made at tarsal/tibia junction. A stick was used to loosen the skin at the cut. The butcher then blew into the carcass to inflate the skin around the body

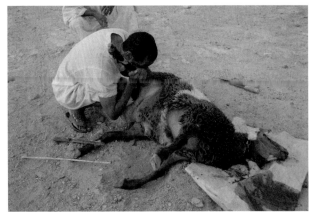

Fig. 12.2 Inflating of the hide

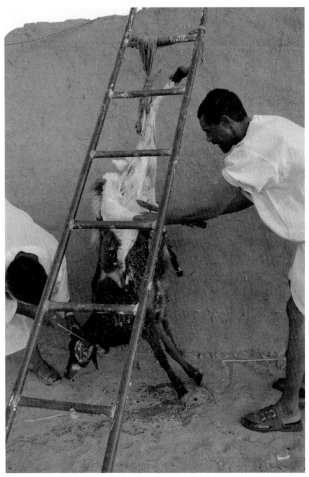

Fig. 12.3 Skinning of the carcass

(Fig. 12.2). Skinning continued with three men working together. One man functioned to hold the legs of the sheep while the other two men continued to skin the hind legs.

Once the hind legs were skinned, the neck area was washed with water and the carcass was hung by the hind legs from a ladder (Fig. 12.3). The head was removed neatly at the occipital condyles by one man while skinning was continued by the second man. The front legs were skinned and cut through at the metacarpal junction and the lower portion removed and tossed away. The remaining portion

Fig. 12.4 Chopping of the front limb into segments

Fig. 12.6 Removal of the ribs from the spine

Fig. 12.5 Chopping of the hind limb into segments

Fig. 12.7 Chopping of the ribs into segments

of the front limb was removed intact with the scapula (Fig. 12.4). Each hind limb was removed at the acetabulum in one section (Fig. 12.5). The pelvis (with tail) was removed as a single section. Each leg and the pelvis were chopped into smaller sections with an axe. The neck was removed using both knives to cut through the meat as well as an axe to chop through the bones. Ribs were removed from the vertebral column, and ethe vertebrae removed from the pelvis in a similar manner in large sections (Figs. 12.6 and 12.7). Each section was placed on large trays to await

Fig. 12.8. Trays of butchered elements

Fig. 12.9 Board utilized for chopping sections of carcass

Fig. 12.10 Chopping of the skinned sheep head

further processing (Fig. 12.8). Testicles were also removed and placed on the trays.

One man continued to remove viscera and wash the carcass while two other men began to further process the portions of meat on the trays. First, the meat was cut from limb sections with a knife, and then the bone portion was chopped into smaller sections with an axe. Chopping was done on a small section of board that had been covered with the skin of the animal (Fig. 12.9). These chopped sections were then further separated with knives. Each of the limb sections was processed in this manner, as was the neck, ribs, pelvis and remaining portions of the vertebral column. Some variability in the butchering sequence does occur in that the slicing of meat from the sections may happen before and/or after chopping. For instance, after removal from the vertebral column, one side of the rib cage was chopped into small pieces, while the other side was sliced into three smaller portions before chopping into smaller sections occurred. The head was skinned and superficially chopped several times with an axe but it was not processed further (Fig. 12.10). Viscera were placed on trays but received no further processing by the men. The total butchering time was approximately one hour.

Only small bone chips were present at the butchering site after processing. The metacarpals and front phalanges, and the metatarsal and hind phalanges, were kept intact and removed to a dumping area, a distance of approximately 100 metres from the butchering area, behind and away from the houses. The dumping area was a shallow surface midden where bones and other refuse were available to dogs and other scavengers. The skin was also discarded in the same area (Fig. 12.11).

Cooking

Although women did not participate in the butchering, they started to select pieces of meat, beginning with the

Fig. 12.11 Midden area for disposal of metacarpals, metatarsals and skin

Fig. 12.13 Making of stew

Fig. 12.12. Cutting of meat segments by women

Fig. 12.14 Bread to be served with the meal

meaty parts and the viscera, as the butchering progressed. The selected portions were immediately brought into the compound kitchen for cooking. Once the butchering was completed, all the trays were collected by women and brought to the kitchen (Fig. 12.12). Men did not participate in any of the preparation of the food. However, the household women were assisted by a large contingent of related and neighbouring women in the preparation of ingredients for the feast which occurred in the women's area of the compound.

Meat was washed with water from a large metal kettle and placed into several pots. Many dishes were prepared with the addition of onion, oil, salt, garlic, pepper, cumin and other spices. Tomato sauce and cinnamon were also added. Several stew combinations were made with meat on the bone, meat without bone, and viscera. The head was prepared whole in a stew (Fig. 12.13). Women did cut portions of the meat into smaller pieces, but when they encountered a section of bone that was too large for the pot it was returned to the men for further processing and chopping with an axe. Interviews revealed that meat was sometimes prepared by roasting without the bone. However, roasting meat with or without bone was not observed in the course of this study.

In addition to meat dishes, stuffed vegetables, including eggplant, were prepared. Bread accompanies every meal (Fig. 12.14). In all households visited, bread was no longer baked in the home on a daily basis, but was purchased from one of the two commercial bakers in town.

Based on these observations, it is expected that sheep/goat faunal remains will show evidence of specific butchering practices. Taphonomic factors will be dominated by chop marks resulting from an axe, cut marks from slicing with a knife, and carnivore marks from discard in open surface middens. In addition, it is expected that butchered remains will be similar in size and may show evidence of having been boiled.

Discard

Faunal remains were systematically collected from five midden areas chosen at random within the town of Delgo. The middens were located near clusters of households and it was assumed that these middens were produced from discard brought to each location from nearby households. Table 12.2 presents a summary of the data collected from the middens which are numbered one through five. A total of 279 bones were examined and a range of butchering marks was photographed in detail.

Consistent with the butchering practices observed, all

Table 12.2 Summary of faunal remains from domestic middens (continued over the next four pages)

Midden Revised	Species	Element	Portion	No.	Butchering	Carnivore	Notes
Midden 1	Bos taurus	Cranial	Indeterminate	4			
Midden 1	Bos taurus	Fused central and 4th tarsal	Complete	1			
Midden 1	Bos taurus	Humerus	Distal epiphysis and 1/4 shaft	1	Chopmarks on shaft and condyle		
Midden 1	Bos taurus	Indeterminate vertebrae		3	Chopmarks		
Midden 1	Bos taurus	Innominate	Indeterminate	3	Chopmarks		Burned and calcined
Midden 1	Bos taurus	Innominate	Acetabulum and portion of ilium	1	Chopmarks		
Midden 1	Bos taurus	LBF	Shaft	8	Chopmarks		
Midden 1	Bos taurus	Mandible with teeth	Horizontal ramus	1			P2, P3 and P4
Midden 1	Bos taurus	Maxilla with teeth		3			
Midden 1	Bos taurus	Metacarpal	Shaft	3	Chopmark at distal end	Present	Longitudinal fracture
Midden 1	Bos taurus	Metatarsal	Proximal epiphysis and 1/4 shaft	3	Chopmark at proximal end		Longitudinal fracture
Midden 1	Bos taurus	Rib	Shaft	11	Chopmarks at both ends	Present	
Midden 1	Bos taurus	Sacrum	Complete	1	Chopmarks		
Midden 1	Bos taurus	Scapula	Blade	3	Chopmarks on both ends and spine		Burned and calcined
Midden 1	Bos taurus	Ulna	Proximal epiphysis	1	Chopmark		
Midden 1	Ovicaprine	Atlas	Anterior portion	1	Chopmark removed posterior half		
Midden 1	Ovicaprine	Cervical vertebrae	No centrum	1	Chopmarks		
Midden 1	Ovicaprine	Cranial	Occipital and frontal bones	3	Chopmarks		
Midden 1	Ovicaprine	Femur	Shaft	1	Chopmarks; cutmarks		
Midden 1	Ovicaprine	Humerus	Distal epiphysis and 1/4 shaft	1	Chopmark through epiphysis		
Midden 1	Ovicaprine	LBF	Shaft	9	Chopmarks		
Midden 1	Ovicaprine	Lumbar vertebrae	Complete	1			
Midden 1	Ovicaprine	Mandible with teeth	Horizontal ramus	7			
Midden 1	Ovicaprine	Mandible without teeth	Condyle and portion of vertical ramus	1	Chopmark on vertical ramus	Present	
Midden 1	Ovicaprine	Maxilla with teeth		3			
Midden 1	Ovicaprine	Radius	Proximal epiphysis and 1/4 shaft	1	Chopmark		
Midden 1	Ovicaprine	Radius	Distal shaft	1	Chopmark at mid shaft		
Midden 1	Ovicaprine	Radius/Ulna	Proximal end and 1/4 shaft	1	Chopmark		Unfused proximal epiphysis

12. Ethnozooarchaeology of butchering practices in the Mahas Region, Sudan

Midden Revised	Species	Element	Portion	No.	Butchering	Carnivore	Notes
Midden 1	Ovicaprine	Radius/Ulna	Proximal end	1	Chopmark on lateral edge		
Midden 1	Ovicaprine	Rib	Shaft	3	Chopmarks at proximal end	Present	
Midden 1	Ovicaprine	Scapula	Blade	3	Chopmarks on both ends		
Midden 1	Ovicaprine	Tibia	Shaft	1	Chopmarks at both ends		
Midden 2	Bos taurus	2nd phalange	Complete	1		Present	
Midden 2	Bos taurus	Astragalus	Near complete	1	Edges removed by chopmarks		
Midden 2	Bos taurus	Cranial	Indeterminate	2	Chopmark		
Midden 2	Bos taurus	Cranial	Occipital condyle	1	Chopmark		
Midden 2	Bos taurus	Fused central and 4th tarsal	Complete	1	Chopmark		
Midden 2	Bos taurus	LBF	Shaft	3			
Midden 2	Bos taurus	LBF	Shaft	3	Chopmark		
Midden 2	Bos taurus	Lumbar vertebrae	Vertebral arch	1	Chopmarks		
Midden 2	Bos taurus	Lumbar vertebrae	Centrum	1	Chopped longitudinally through centrum		
Midden 2	Bos taurus	Mandible with teeth	Horizontal ramus	1	Chopmark removed ramus posterior to M2		
Midden 2	Bos taurus	Metacarpal	Shaft	3	Chopmark midshaft		Longitudinal fracture
Midden 2	Bos taurus	Metatarsal	Proximal epiphysis and 1/2 shaft	1	Chopmarks midshaft		
Midden 2	Bos taurus	Rib	Shaft	7	Chopmarks on both ends; cutmarks present	Present	
Midden 2	Bos taurus	Scapula	Blade	1	Chopmarks on both ends		
Midden 2	Bos taurus	Scapula	Glenoid fossa	1	Chopmark		
Midden 2	Bos taurus	Thoracic vertebrae	Vertebral arch	1	Chopmarks		
Midden 2	Bos taurus	Ulna	Proximal shaft	1	Chopmark on proximal end; cutmarks		
Midden 2	Canis familiaris	Mandible without teeth	Complete	1			
Midden 2	Equus asinus	Metapodial	Distal shaft	1			
Midden 2	Ovicaprine	Caudal vertebrae	Centrum	1	Chopmarks removed transverse processes		
Midden 2	Ovicaprine	Femur	Shaft	1	Chopmark midshaft		
Midden 2	Ovicaprine	Indeterminate vertebrae	Vertebral arch	2	Chopmarks		Fused
Midden 2	Ovicaprine	Mandible with teeth	Horizontal ramus	3	Chopmarks		
Midden 2	Ovicaprine	Rib	Shaft	3	Chopmark		Calcined

Midden Revised	Species	Element	Portion	No.	Butchering	Carnivore	Notes
Midden 2	Ovicaprine	Scapula	Blade	1	Chopmarks on both ends		
Midden 2	Ovicaprine	Tibia	Distal epiphysis and 1/2 shaft	1			Longitudinal fracture
Midden 3	Bos taurus	1st Phalange	Complete	4	Chopmark		
Midden 3	Bos taurus	2nd phalange	Complete	1	Chopmark on distal end		
Midden 3	Bos taurus	3rd Phalange	Complete	1			
Midden 3	Bos taurus	Indeterminate vertebrae	Centrum	1			
Midden 3	Bos taurus	LBF	Shaft	4	Chopmarks		
Midden 3	Bos taurus	Mandible without teeth	Mandibular condyle	1	Chopmarks		Longitudinal fracture
Midden 3	Bos taurus	Metacarpal	Proximal epiphysis and 1/2 shaft	2	Chopmark at end of shaft		Longitudinal fracture
Midden 3	Bos taurus	Metatarsal	Distal shaft; unfused epiphysis	1	Chopmark at proximal end		
Midden 3	Bos taurus	Rib	Shaft	14	Chopmarks on both ends	Present - rodent	
Midden 3	Bos taurus	Thoracic vertebrae	Distal epiphysis	1	Cutmarks present		
Midden 3	Ovicaprine	Femur	Shaft	1	Chopmarks on both ends		Longitudinal fracture
Midden 3	Ovicaprine	LBF	Shaft	5	Chopmarks		
Midden 3	Ovicaprine	Mandible without teeth	Mandibular condyle	1	Chopmark on vertical ramus	Present	
Midden 3	Ovicaprine	Maxilla with teeth	P4, M1 and M2 present	1			Epiphyseal fusion still visible
Midden 3	Ovicaprine	Radius/Ulna	Distal epiphysis and 1/2 shaft	1	Chopmark on shaft		
Midden 3	Ovicaprine	Rib	Shaft	5	Chopmark proximal end		
Midden 3	Ovicaprine	Tibia	Distal epiphysis and 1/4 shaft	1	Chopmark on shaft		
Midden 4	Bos taurus	1st Phalange	Complete	3	Chopmark	Present	
Midden 4	Bos taurus	3rd Phalange	Complete	1			
Midden 4	Bos taurus	Astragalus	Complete	1	Chopmarks at both ends		
Midden 4	Bos taurus	Atlas	Lateral articular surface for occipital condyle	1	Chopmarks split vertebrae medially		
Midden 4	Bos taurus	Cervical vertebrae	Centrum	1	Chopmark vertically through body		Unfused
Midden 4	Bos taurus	Cunieform	Complete	1	Chopmarks and cutmarks present		
Midden 4	Bos taurus	Fused 2nd and 3rd tarsal	Complete	1			
Midden 4	Bos taurus	Indeterminate		1			

12. *Ethnozooarchaeology of butchering practices in the Mahas Region, Sudan*   115

Midden Revised	Species	Element	Portion	No.	Butchering	Carnivore	Notes
Midden 4	*Bos taurus*	Innominate	Illium	2	Chopmarks	Present	
Midden 4	*Bos taurus*	Innominate	Indeterminate	2	Chopmarks		
Midden 4	*Bos taurus*	LBF	Shaft	2	Chopmark		
Midden 4	*Bos taurus*	Lumbar vertebrae	Transverse process	4	Chopmark	Present	
Midden 4	*Bos taurus*	Metapodial	Unfused distal epiphysis	1			
Midden 4	*Bos taurus*	Metatarsal	Shaft	2	Chopmark at proximal end	Present	
Midden 4	*Bos taurus*	Radius	Posterior shaft	1	Chopmark on distal end		
Midden 4	*Bos taurus*	Rib	Shaft	6	Cutmarks and chopmarks	Present	
Midden 4	*Bos taurus*	Thoracic vertebrae	Dorsal process	1	Chopmarks at base of process and proximal end	Present	
Midden 4	*Bos taurus*	Vertebrae		2	Chopmark		
Midden 4	Chicken	Femur	Distal epiphysis and 3/4 shaft	1	Cutmarks above distal epiphysis		
Midden 4	Chicken	Innominate		1			
Midden 4	Ind. Pisces	Indeterminate	Indeterminate	1			
Midden 4	Ind. Pisces	Vertebrae	Centrum	1			5 articulated vertebrae
Midden 4	Ovicaprine	Femur	Proximal shaft	2	Chopmarks		
Midden 4	Ovicaprine	Humerus	Distal epiphysis and 1/4 shaft	3	Chopmarks		
Midden 4	Ovicaprine	Innominate	Acetabulum, illium and ischium	1	Chopmark on ilium and ischium	Present	
Midden 4	Ovicaprine	Innominate	Acetabulum and pubis	1	Chopmark		
Midden 4	Ovicaprine	LBF	Shaft	3			
Midden 4	Ovicaprine	Lumbar vertebrae	Complete	1	Chopmark in anterior end	Present	2 articulated vertebrae
Midden 4	Ovicaprine	Mandible with teeth	Horizontal ramus	1	Chopmark behind M2	Present	M2 damaged
Midden 4	Ovicaprine	Rib	Shaft	7			
Midden 4	Ovicaprine	Sacrum	Complete	1	Chopped on both sides and at distal end		
Midden 4	Ovicaprine	Scapula	Proximal end	1	Chopmarks		
Midden 5	*Bos taurus*	1st Phalange	Complete	2	Chopmarks		
Midden 5	*Bos taurus*	2nd phalange	Complete	1			
Midden 5	*Bos taurus*	3rd Phalange	Complete	1			
Midden 5	*Bos taurus*	Calcaneous	Proximal epiphysis	1	Chopmarks and cutmarks present		
Midden 5	*Bos taurus*	Cranial	Zygomatic arch	1			Burned
Midden 5	*Bos taurus*	Cranial	Indeterminate	1			Boiled

Midden Revised	Species	Element	Portion	No.	Butchering	Carnivore	Notes
Midden 5	Bos taurus	Femur	Distal epiphysis	1	Chopmarks		
Midden 5	Bos taurus	Humerus	Distal epiphysis	1	Removed by chopping		Heavy weathering
Midden 5	Bos taurus	Indeterminate tarsal	Complete	1	Cutmarks present		
Midden 5	Bos taurus	Indeterminate vertebrae		2	Chopmarks		
Midden 5	Bos taurus	Innominate	Indeterminate	1	Chopmarks		Burned/calcined
Midden 5	Bos taurus	LBF	Shaft	1		Present	
Midden 5	Bos taurus	Mandible with teeth		2	Chopmarks		
Midden 5	Bos taurus	Maxilla with teeth		2			
Midden 5	Bos taurus	Metacarpal	Distal epiphysis and 1/4 shaft	1	Chopmark at epiphysial junction		
Midden 5	Bos taurus	Metatarsal	Proximal epiphysis and 1/2 shaft	1	Chopmark midshaft		
Midden 5	Bos taurus	Radius	Proximal epiphysis	1			Boiled
Midden 5	Bos taurus	Rib	Shaft	11	Chopmarks		
Midden 5	Bos taurus	Scapula	Glenoid fossa	1	Chopmark through fossa		
Midden 5	Bos taurus	Scapula	Spine	1	Chopmarks		
Midden 5	Bos taurus	Thoracic vertebrae	Transverse process	1	Chopmarks		
Midden 5	Bos taurus	Thoracic vertebrae	Transverse process	1	Chopmarks		
Midden 5	Bos taurus	Tibia	Distal epiphysis	1	Chopmarks		
Midden 5	Bos taurus	Ulna	Proximal epiphysis	1	Chopmarks		
Midden 5	Equus asinus	Metacarpal	Complete	1			
Midden 5	Equus asinus	Radius/Ulna	Complete	1			
Midden 5	Ovicaprine	Cervical vertebrae	Complete	2	Chopmarks		
Midden 5	Ovicaprine	Maxilla with teeth	P3, P4, M1 and M2 present	1			
Midden 5	Ovicaprine	Radius	Shaft	1	Chopmarks midshaft		
Midden 5	Ovicaprine	Radius/Ulna	Shaft	1	Chopmark proximal end		
Midden 5	Ovicaprine	Thoracic vertebrae	Vertebral arch	1			

Fig. 12.15 Butchered sheep/goat rib fragments

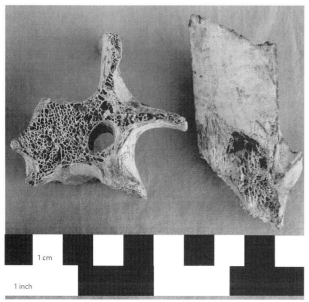

Fig. 12.16 Butchered sheep/goat vertebrae fragments

remains collected from the middens were smaller than 10cm in maximum length (Figs. 12.15 and 12.16). Both cattle and sheep/goat remains are present and other species such as chicken, donkey and dog also occur; some fish was also collected. Fish were never observed being sold in the Delgo market and either these fish were caught in the Nile River on an intermittent basis or were purchased in more distant markets. Donkey remains show no evidence of butchering. Their presence in the assemblages is not surprising due to their prevalence as transportation and pack animals in the community. Carnivore gnawing was present on a proportion (16%) of the remains. As many feral dogs are present and uncontrolled within the community, they are often foraging among the garbage areas. A single dog mandible was recovered from the assemblage.

Seventy-one percent of the collected faunal assemblage shows butchering marks which are predominantly chop marks from axes. By species, 77% of the cattle remains show butchering evidence as do 54% of the small stock remains. The discard patterns observed in the middens conform strongly with expectations based on the observed butchery practices. Cattle are slaughtered by a local professional butcher and the commercial midden included only mandibles (Arnold 2008, in prep.). The domestic middens contained a large number of cattle metacarpals, metatarsals and phalanges. Informants stated that these cattle bones are purchased from the butcher for making soup for daily meals. Not surprisingly, cattle bone elements in domestic contexts also show a high proportion of longitudinal fractures for marrow extraction, which are for instance visible on 44% of the metacarpals and 50% of the metatarsals.

The minimal amount of burned bones recovered from the domestic midden assemblages further indicates that roasting for the preparation of meat for consumption is infrequent. The majority of burned remains in the midden samples were likely caused by occasional and minimal attempts to burn garbage within the village, a practice witnessed on several occasions over the course of the study.

Discussion

The call for the ethnoarchaeological examination of cooking processes (Gifford-Gonzalez 1993) emphasizes the need to evaluate time and energy costs for processing against the cost of transporting meat parts to locations of consumption. This type of research has been profitable in hunter-gatherer contexts and should be applied to sedentary and more urban contexts where transport costs are minimal.

Energetic costs of the various cooking and processing techniques, as well as their nutritive benefits, should also be considered (Gifford-Gonzalez 1993). Delgo families do not consume meat on a regular basis and the slaughter of domestically owned livestock (excepting occasional eating of chickens and pigeons) is reserved for special occasions. There is a low density of livestock throughout the community which makes the efficient processing of meat resources a priority. Complete utilization of an animal in order to extract all available nutrients is evident in the observed processing and cooking practices. In addition, the two feasting events directly observed in the study were able to feed a large group of people. One of the feasts was to celebrate the homecoming of a male member of the household from work overseas and included approximately 50 guests. The second feast was a naming ceremony for a newborn child and included approximately 150 guests. Each of these feasts was provisioned by the slaughtering of one sheep.

As discussed above, the animal for each feast was first butchered into large sections (*e.g.* the front limb, the hind limb, ribs and vertebrae). These were then chopped with an axe into small sections which were further reduced through more precise cutting with knives. The resulting butchering pattern is one in which all elements are removed from the body, chopped with an axe to break the bones and then cut with knives into manageable (pot-sized) pieces. A similar pattern was also recorded in the commercial context where a cow was butchered. The same "pot-sized" pieces are produced, and the animal was distributed throughout the community by women purchasing small portions of meat for their respective households (often in cuts of viscera or of meat with or without bone).

Based on these observations it is argued here that women have an effect on men's butchering practices. However, butchering practices are usually interpreted by archaeologists as being constituted in decisions made by men in consideration of transport (Binford 1978; Bunn *et al.* 1988; Gifford-Gonzalez 1991; 1993; O'Connell *et al.* 1988; Yellen 1977). While men are solely responsible for the act of butchering in Delgo, it is women's considerations in cooking practices that dictate men's decisions in how they process an animal.

Gender and gender relations are generally recognized to be reproduced and transformed through feasting, although women's contributions and ambitions in feasting are not well studied (Dietler and Hayden 2001; Gero 1992). Of interest here is the gendered asymmetry in terms of labour and benefits (Dietler and Hayden 2001). The gendered division of labour in the feasts presented here involved males as the hosts and suppliers of the feast, and women's cookery practices as influencing how animals were butchered and then prepared. Kifleyesus (2002) has examined the social and symbolic function of food among the Muslim Argobba of Ethiopia. He examines a wide range of factors including food politics and aesthetics, gendered meanings behind diet and cuisine, socialization in foodways and how meals construct both ethnic identity and social boundaries. The gendered division of labour between butchering and cooking is summarized as "men slaughter and slice and women stew and spice" (Kifleyesus 2002, 262). Kifleyesus (2002, 262) concludes that the act of preparing and serving a meal by women is an act of submission and "seems to place the Argobba woman at the bottom of the social hierarchy". Benkheira (1999) notes similar patterns among the Islamic people of Maghreb in North Africa. However, other researchers recognize (Appadurai 1981; Benkheira 1999; Kifleyesus 2002) that women can generate power through their control over preparing and serving food. In this study it is apparent that women's contribution to feasts goes beyond cooking and serving animal products that are given to them by men in their household. Instead, men's and women's roles in animal management, butchering and cookery are integrated processes in which women play a significant role. Obviously the full extent of the social and power relations in the type of communal feasting recorded in this short-term investigation has not been realized. However, an important avenue for further investigation should be to determine more fully the perceptions and attitudes of feast participants, such as the potential for women to shame their husbands by not preparing meat "properly". Alternatively, women in the host household were also joined and helped by many female guests and both women and men participate in the feast, but they eat in separate spatial contexts within the host's compound. The potential for women to compete for status with one another through their individual knowledge, skill, and innovation in culinary practices is highly probable.

The examination of faunal remains from midden areas in Delgo can perhaps provide a small but useful contribution to the study of feasting within the archaeological record. There is an obvious link between butchering practices recorded in both the domestic setting and the commercial setting, and the bone debris collected throughout the town of Delgo. The negligible amount of bone remains at butchering sites produced by domestic butchery make archaeological identification of butchering locales improbable. However, it is anticipated that an archaeological assemblage that resembles the butchering pattern observed here would display the following characteristics and may indicate similar processing techniques. Three general characteristics are suggested:

- Remains with a high percentage of chopmarks (>60%)
- The majority of remains <10cm in length
- The absence of small stock metapodials and phalanges.

It is suggested that an assemblage with these criteria reflect an urban setting with infrequent domestic consumption of livestock, a combination of butchering practices both in the domestic and professional setting, full utilization of the carcass for maximum nutritive benefit, and a processing style that reflects a dominant cookery practice of stews and soups. The full utilization of the carcass and the preparation of stews are reflective of the efficient use of small amounts of meat in preparing feasts for large groups of people. Butchering patterns are driven by cooking practices in which both men and women participate.

Conclusions

Gifford-Gonzalez (1993) called for more investigation of the relationship between culinary processes and butchering practices as one avenue for addressing gender bias in zooarchaeology. While originally applied to hunter-gatherer contexts, this investigation has expanded the application of these fundamental ideals to an urban context in Sudan with informative results. Within this investigation, it has been shown that cooking practices strongly influence butchering decisions in carcass preparation. A strong avenue for further research is the social and cultural perceptions of these practices.

The butchery patterns observed on discarded bones found in middens accurately reflect the butchering practices observed directly in the preparation of feasts in domestic contexts. Based on these observations, a set of criteria were proposed that allow for the identification of similar butchering practices and consumption patterns in archaeological contexts. This type of data may be useful to archaeological investigations of feasting in other contexts.

Acknowledgements

Funding for this portion of the project was provided by a URGC starter grant from the University of Calgary (grant holder Dr. Diane Lyons). Thanks are also extended to Dr. A. C. D'Andrea (Simon Fraser University). Valuable comments and revisions were made by Andrew Reid and Umberto Albarella.

References

Appadurai, A. (1981) Gastro-Politics in Hindu South Asia. *American Ethnologist* 8(3), 494–511.

Benkheira, M. H. (1999) Binding and separating: the ritual function of meat in the Islamic world. *L'Homme* 152, 89–114

Binford. L. (1978) *Nunamiut Ethnoarchaeology*. New York, Academic Press.

Bravo-Baumann, H. (2000) *Capitalisation of experiences on the contribution of livestock projects to gender issues*. Working Document. Bern, Swiss Agency for Development and Cooperation.

Bunn, H. T., Bartram, L. E. and Kroll, E. M. (1988) Variability in bone assemblage formation from Hadza hunting, scavenging, and carcass processing. *Journal of Anthropological Archaeology* 7, 412–457.

Dahl, G. and A. Hjort (1976) *Having herds. Pastoral herd growth and household economy*. Stockholm Studies in Social Anthropology 2. Department of Social Anthropology, University of Stockholm. Liber Tryck, Stockholm.

David, N. and Kramer, C. (eds.) (2001) *Ethnoarchaeology in Action*. Cambridge, Cambridge University Press.

Dietler, M. and Hayden, B. (eds.) (2001) *Feasts: archaeological and ethnographic perspectives on food, politics and power*. Washington, Smithsonian Institution Press.

FAO (1995). Women, Agriculture and Rural Development in the Near East: Findings of an FAO Study. Extracted from *Women, Agriculture and Rural Development: A Synthesis Report of the Near East Region*.

Gero, J. (1992) Feast and Females: Gender Ideology and Political Meals. *Norwegian Archaeological Review* 25, 1–16.

Gifford-Gonzalez, D. (1991) Bones are not enough analogues, knowledge and interpretive strategies in zooarchaeology. *Journal of Anthropological Archaeology* 10, 215–254.

Gifford-Gonzalez, D. (1993) Gaps in the zooarchaeological analyses of butchering: is gender an issue? In J. Hudson (ed.) *From Bones to Behaviour: Ethnoarchaeological and experimental contributions to the interpretation of faunal remains*. Center for Archaeological Investigations, Occasional Paper No. 21., 181–199. Carbondale, Southern Illinois University.

Kifleyesus, A. (2002). Muslims and Meals: The Social and Symbolic Function of Foods in Changing Socio-Economic Environments. *Africa: Journal of the International African Institute* 72(2), 245–276.

Monod, T. (1975) Introduction. In Monod, T. (ed.) *Pastoralism in Tropical Africa*, 8–98. International African Institute London, Oxford University Press.

Niamir, M. (1982) Report on animal husbandry among the Nqok Dinka of the Sudan: Integrated rural development Project, Ayei, South Kordofan, Sudan. *HIID Rural Development Studies* Cambridge, Harvard University.

O'Connell, J. F., Hawkes, K. and Blurton Jones, N. (1988) Hadza hunting, butchering and bone transport practices and their archaeological implications. *Journal of Anthropological Research* 44, 113–161.

Yellen, J. E. (1977) Cultural patterning in faunal remains: evidence from the !Kung Bushmen. In D. Ingersoll, J. E. Yellen, and W. MacDonald (eds.) *Experimental Archaeology*, 271–331. New York, Columbia University Press.

Husbandry and Herding

13. Social principles of Andean camelid pastoralism and archaeological interpretations

Penelope Dransart

This paper considers some social principles associated with pastoralism. Selected for discussion here are principles concerning networks of ownership, patterns of access to pasture grounds, and interactions between the social organization of the human group and that of their herd animals. On the basis of an ethnographically observed community of herders of llamas, alpacas and sheep in Isluga, northern Chile, the paper tracks changes in the social organization of a pastoral way of life from the 1980s to the beginning of the twenty-first century. It is argued that herding practices are based on social principles which, ultimately, make an impact on biological processes of domestication. The use of ethnographic data in archaeological interpretations can have the effect of conveying the notion of a timeless, unchanging present. Therefore this paper invites archaeologists to consider carefully their use of ethnographic analogy by taking into account the changing social contexts which generated the data.

Keywords: South American camelids, social principles of pastoralism, Isluga, ethnography, ethnographic analogy

Introduction

In this paper, I consider the engagement between human beings and herd animals in a way of life which derives its character from the quality of the social interaction between humans and South American camelids. The emphasis here is on certain social principles associated with pastoralism rather than the biological processes of domestication. It is my contention that practices emerging from such social principles ultimately have an effect on the domesticating process of herd animals.

The etymology of the term "pastoralism" is derived from the Latin "to feed". It is interesting to place this term in parallel with the Aymara term *uywa*, which is often understood to be the equivalent of the English 'domesticated animal'. Literally, an *uywa* is a "cared for" animal, the verbal form being *uywaña*, "to rear [animals], to breed, to nourish or foster" (Apaza Suca *et al.* 1984, 238; Mamani M. 2002, 164). This term cuts across the aspects of "rearing" and "breeding", which Ingold (1986, 168) regarded as insufficient for identifying the essential characteristics of pastoralism as a subsistence economy, and the aspect of protecting herd animals. Protection is a practical activity which forms part of what Ingold sees as pastoralism's defining characteristic: "the social appropriation, by persons or groups, of successive generations of living animals" (Ingold 1980, 133).

At one time it was thought that pastoralism did not develop as a "dominant economy" anywhere in the Americas, and that it was only with the introduction of sheep and horses to the North American Southwest that the Navajo were able to adopt a pastoralist way of life (Forde 1934, 394). The publication of Jorge Flores Ochoa's ethnography of the alpaca-herding community of Paratía, in the highlands of southern Peru, first in Spanish, then in English, overturned Forde's argument that Andean camelid herding constituted "an auxiliary and integral part of a developed agricultural and sedentary civilization" (Forde 1934, 394; Flores Ochoa 1968; 1979). In the 1960s, Flores Ochoa did not have access to the growing body of detailed zooarchaeological findings which have been published subsequently. This literature indicates that the herding of camelids arose by 5000 BP and it has been reviewed by Mengoni Goñalons and Yacobaccio (2006) and Wheeler *et al.* (2006).

In the ethnographic literature, pastoralism is a subsistence strategy in which the caring for herd animals is based on the social principle of ownership. While access to live animals (over which individual human owners claim ownership) is divided, access to pasture grounds is held in common. Individual animals are typically herded in collective family units (Caro 1985). In contrast with a hunting economy, people assumed responsibility for the care of their herd animals. They broke with the principle of having unrestricted access to herbivores that they

	Access to animals	**Access to land**
Hunting	common	common
Pastoralism	divided	common
Ranching	divided	divided

Table 13.1 The distribution of access to animals and land (after Ingold 1980, 5)

might hunt. A deep commitment is necessary on the part of the owners of herbivores and it is the quality of this relationship that characterizes pastoralism as a way of life (Flores Ochoa 1979, 8). Pastoralists do not normally grow foodstuffs to store and give to their animals at a later stage in the annual cycle. An important characteristic of a pastoralist economy is that the herders take their animals to different pastures on a seasonal basis in order to ensure access to adequate sources of food. The tripartite relationships between human beings, herd animals and pasture lands are presented in a grid form in Ingold (1980, 5) (Table 13.1). In his book, *Hunters, pastoralists and ranchers*, he adopted an evolutionary approach to address the question of what happened in subarctic regions in times of food shortage among hunters of reindeer. The transformations which he tracked proceeded from hunting to pastoralism to ranching. He characterized hunting and ranching by their predatoriness, in contrast to pastoralism, which is "protective" (Ingold 1980, 2). His own fieldwork experience was amongst ranchers of reindeer (Ingold 1980, 20).

My fieldwork, in contrast, has been with herders of llamas, alpacas and sheep in Isluga, in the highlands of the far north of Chile. Using Ingold's grid, they are classified as pastoralists. Generally speaking, pasture lands are held in common in Isluga and people herd their own animals in family units. Their herd animals play an important role as a focus of ritual activity, which is one of the criteria for defining pastoralism employed by Flores Ochoa (1979, 8). A brief description of Isluga's subsistence economy might imply a somewhat static, "traditional" society. However, human and herd demographics are constantly changing. In recent years the emergence of a ranching economy in Isluga has not occurred. Instead, the social transformations occurring as human migration makes an increasing impact on the way of life have been following a different trajectory, some aspects of which I will present below.

In this paper, I address the social principles of pastoralism and the importance of understanding those principles if we wish to examine archaeological questions through the study of human-herd animal relations. A modified comparative perspective is used in which I examine pastoralism as practised in Isluga and the social changes that have occurred as herders attempt to maintain continuity in their pastoral economy through time. This material is considered in the light of some of the transformations that have occurred in the social conditions experienced by pastoralists in other parts of the world (sub-arctic Eurasia and the Arabian Peninsula). My aim here is to explore ways of perceiving the diversity of pastoral economies, not just between the Andes and other parts of the world, but also within the Andes and between past and present. Tim Ingold (1980, 7) pointed out that "social evolution does not consist in the cumulative record of cultural innovations, but involves a series of transformations in the very conditions to which they emerge as functional responses".

The social principles with which I am concerned here are as follows:

1. *networks of ownership* in which the herders establish dependencies between human members of the group through the individual ownership of animals as property;
2. *social patterns of access to pasture lands*. This aspect has to do with maintenance of habitat and control of territory. Herders have to maintain both herds of herbivores as well as pasture grounds, which are usually communal;
3. *interactions between the social organization of the human owners and that of their herd animals*. This aspect has to do with the cultural behaviour and the behavioural ecology of both human beings and the camelids or sheep they herd.

Pastoralism in Isluga and Its Social Transformations

The ethnographic material on which this paper is based derives from my fieldwork in Isluga, northern Chile, which I have conducted since 1986. Until the late nineteenth-century, Isluga formed part of the southernmost part of Peruvian national territory. The Chilean occupation of 1879 led to a proposed plebiscite in 1929. Isluga, as part of the Province of Tarapaca, then became incorporated into Chilean territory. Isluga is a bilingual Aymara-Spanish speaking community of some 2000 persons whose territory is bounded by the Salar de Surire on the north and the Salar de Coipasa on the south, adjacent to the border with Bolivia. Administratively, it is combined with Cariquima in what is now called the I Región of Chile; this history has been explored by Ortega Perrier (1998). In order to place my ethnographic work in a historical context, it is worth explaining briefly the background.

During the first twenty years of the Spanish occupation of the Andes in the sixteenth century, the colonizers initiated attempts to resettle Andean populations in small towns or larger villages. Between 1569 and 1581, Viceroy Toledo instigated a sustained project of resettlement in the Viceroyalty of Peru (which covered a greater geographical extent than the nation state of today), making it possible to collect revenue and information (MacCormack 1991, 140). The grid-plan layout of resettled communities or *reducciones* (from the Spanish verb *reducir*, "to reduce") was intended to denote the civility of the resettled inhabitants and their reception of instruction in the Christian faith. Valerie Fraser (1990, 41) pointed out the interdependence between "the civic, the civil and the Christian" which prevailed in the later sixteenth century. However Sabine MacCormack (1991, 141) observed that the formal, Hispanic appearance of *reducciones* did not necessarily mean that their Andean inhabitants held Christian beliefs and/or observed Christian conduct.

It is obvious that permanent settlement in a *reducción* under the control of one of the missionary orders which operated in the Andes would have undermined the nomadic way of life in Isluga. A compromise was evidently established in the form of a town called Islug marka, which has a central church surrounded by houses on a grid-plan organized in a U-shaped layout (Martínez 1976). It is situated between the two moieties (Araxsaya, the "upper", and Manqhasaya the "lower" or "inner" moiety). People from both moieties congregated in Islug marka for rainy season religious events in the Christian calendar: the festivities associated with the All Souls at the beginning of November, when the rains are commencing; the month of festivals that were formerly observed in December, culminating in St. Thomas the Apostle on 21 December; and Carnival, an event with a moveable date which begins on the Saturday before Ash Wednesday at the end of the rainy season. The observation of these occasions has resulted in patterns of congregation and dispersal which make a herding way of life possible. Dispersion is most marked during the dry or windy season between April and October. A different cycle of events of a more-or-less Christian appearance took place in the scattered communities located throughout Isluga territory, with access to different pasture grounds.

The cyclical interrelationship between ritual and subsistence activities characterized the way of life in Isluga when I first arrived in the mid-1980s (Dransart 2002, 56–58). Therefore, the brief account of pastoralism as practised in Isluga from that time to the beginning of the twenty-first century presented here includes information that is intended to illustrate the interrelationship between ritual observance and subsistence activities. Pastoral production results in a range of resource procurement strategies which also incorporate herd animals in a wider sphere comprising the cultural life of the community.

A pragmatic assumption implicit in many zooarchaeological studies is that humans are at the top of the food chain and that they exploit animals in order to eat them. An example in the archaeological literature which counters such a supposition and which examines the social principles of monastic asceticism is provided by O'Sullivan (2001). She used evidence supplied by the faunal record of early medieval Lindisfarne to consider dietary restrictions on meat and also the need for the monastery to provide a steady supply of calf skins for the *scriptorium*. However, a more characteristic attitude towards consumption in the literature is narrowly concerned with food. Milner and Miracle (2002, 1) introduced an edited volume on patterns of consumption with these words: "It is fascinating to see how attitudes towards food change through time". They evidently did not invite their contributors to considers other aspects of consumption. A critique of such an attitude towards non-human species is explored in Dransart (2002, 12–14). Camelids are essential in providing meat, bones for tools, hides and fleece without which their owners cannot live in the extreme climate of Isluga. (Many of their herding grounds are at 4,000m above sea level and higher.) Herders rise before dawn, when it is bitterly cold, and go out in dangerous conditions in order to herd their camelids, especially during the birthing season when the newly-born llamas and alpacas are at risk from predatory foxes. The birth season tends to takes place between December and March, and lactation continues until about the end of July in Isluga, by which time the rains have ceased and pasture has become scarce (Dransart 2002, 58).

Lightning in a steppe-like terrain is a hazard that endangers both herders and their herd animals. At the onset of the rains in November, people used to associate Saint Andrew with rain and lightning. Effigies of this saint are not known in Isluga, or in the neighbouring province of Carangas in Bolivia, where Monast (1972, 74, 80) called him "a rain devil". If the rains arrived late, herders in the Isluga community in which my fieldwork was based sacrificed a llama in front of the church and invoked Saint Andrew as "Lord of the Rains". They danced in zig-zagging lines to mimic the lightning of the saint and carried white flags, the colour of the rain-bearing clouds as well as that of the llamas claimed by the saint. Other sometimes vengeful Christian saints associated with lightning in Isluga are Saint Thomas, Saint Barbara and Saint James. The point here is that herding activities in a steppe-like environment frequently expose people and herd animals to the risk of death from lightning. Although they do not now carry out the ceremony to Saint Andrew in front of the church, Isluga people continue to observe the restriction against eating the meat of camelids struck by lightning, or spinning the fleece, for fear of boils erupting from one's skin (Dransart 2002, 54–55).

Beyond the taboo against eating such camelids, herders have the ultimate power to decide when to cull their herd animals. However, in their day-to-day organization of herding activities Isluga people work with the social organization of the herd rather than imposing their will unilaterally on the animals. Isluga camelids obey two verbal commands: *kuti* means "turn round" and *piska* means "keep going". The young animals in a herd learn these commands from the older llamas or alpacas, by following the lead of the guide animals. The herders do not instruct their animals as one would instruct a dog (Dransart forthcoming).

In return for caring for herd animals, the herder takes from the animals the materials necessary for sustaining human life in a terrain characterized by its high altitude and unpredictable weather conditions. I understand this

Fig 13.1 Alpacas (foreground) and llamas (at rear) before being ritually "dressed" with brightly dyed fleece and tassels in the wayñu ceremony. Isluga, northern Chile. Photograph by P. Dransart.

Fig. 13.2 The ceremonial "dressing" of female llamas with brightly dyed fleece and tassels during the wayñu ceremony to the accompaniment of singing. Isluga, northern Chile. Photograph by P. Dransart.

engagement with herd animals to serve as a means for bringing each generation of camelid into the social order of the human community to which it belongs. The process is most vividly expressed in the *wayñu* ceremony which is intended to promote the generation of herd animals in parallel with that of their human owners (Dransart 2002, 96–97).

The *wayñu*, or the marking ceremony of the herd animals, is an elaborate undertaking which takes place for camelids over several days between the beginning of January and Ash Wednesday (Fig. 13.1). It is observed by individual families in the communities that are scattered throughout Isluga. The *wayñu* of sheep either takes place at this point in the calendar or later in the season, at the time of the June solstice. A discussion of the event is useful in that it helps to highlight the social principles which I outlined above. An important part of the proceedings is devoted to the ritual investiture of the camelids with ear tassels and brightly dyed fleece (Fig. 13.2). In the case of animals approaching sexual maturity, the owners cut marks in the ears of the animal to make them bleed. Herders state explicitly that they observe the *wayñu* to enhance the fertility of their herd animals. The social principles with which I am concerned in this paper can be seen in the following connections with the *wayñu*.

Networks of ownership. A characteristic of Isluga herding practice concerns the diverging as opposed to the unilineal devolution of property. Heritable property, of which herd animals constitute an important component, is given by parents to children of both sexes. This diverging devolution of property is accompanied by bilateral kinship reckoning in which kinship is reckoned from both parents. A parent gives to his or her children from a young age (both boys and girls) female llamas, alpacas and sheep in recognition of their contribution towards caring for the family's herds. Parents often use the *wayñu* ceremony to make a public gift of a female llama to one of their children. More rarely, a child who is a successful herder will gift a llama to a parent.

Social patterns of access to pasture grounds. During the *wayñu*, herders may go to the different pasture grounds used by their herds in order to make libations of alcohol. Specimens of pasture are also placed on the ritual table outside the corral in which the main events take place during the ceremony.

Interactions between the social organization of the human owners and that of the herd animals. During the *wayñu*, the herders burn *parina* (flamingo) feathers in the hope that their herd will behave like the birds, which frequent lakes in Isluga territory, forming close, cohesive flocks. Camelids are not allowed to go feral in Isluga and herders seek out individual animals which stray from the herd. This attention to camelid welfare can result in family members spending considerable time looking for their animals, and frequent offenders are likely to be culled at a fairly young age.

The ethnographic detail reported above is provided by a society whose subsistence strategy is dominated by the herding of llamas, alpacas, sheep and donkeys, supplemented by the cultivation of potatoes and quinua, a grain which can be grown at altitudes up to 4000m above sea level. Within recorded history, Isluga people have neither used the indigenous foot plough or the traction plough, which the Spanish introduced to many parts of the Andes during the Colonial period. Instead, the cultivation of potatoes and quinua has been assisted by the use of a hand-held digging tool known as *lampa* (Donkin 1979, 9). In the first few years of the twenty-first century people in the moiety of Manqhasaya began to acquire tractors in the lowermost zone of Isluga territory, which is immediately adjacent to the international frontier with Bolivia, in an effort to increase the production of quinua.

In keeping with an economy that is predominantly pastoralist, the demands of the herding cycle tend to drive the annual round of events. Different nomadic cycles take place within Isluga territory during the rainy season (November to March) over distances which can be achieved in a day. A longer form of movement takes place from Isluga westward to the valleys of the precordillera during the windy season (April to the onset of the rains in the highlands), if the rains there make the arduous journey worthwhile. This journey takes from three to five days (Dransart 2002, 45). Such nomadic movements have been classified as, respectively, horizontal (taking place in the same altitudinal zone) and vertical (crossing from altitudinal zone to another to make seasonal use of pasture and water) (Ingold 1986, 182–183).

Isluga herders look at pasture grounds with a practised eye whenever they travel and they greatly admire the greenness of pastures which have been revived by rainfall. They often refer to plants which they claim their own animals find appetizing. Sheep are said to favour *lampaya*, a shrub which grows in arid conditions throughout Isluga territory; it is also recommended for human consumption in the form of a herbal tea to treat coughs. Other plants are available more locally and different herds learn to appreciate the differences in the availability of pasture. One herder explained to me that her llamas liked eating *sura*, a plant which grows in the wet pastures (*bofedal*) of her natal community but not in those of her husband, the two places to which she habitually took her herd. Hence networks of ownership of camelids are closely associated with social patterns of access to pasture grounds.

People herd their camelids and sheep on the pasture grounds next to their community of origin. By declaring that her llamas were accustomed to eating *sura*, the herder in the previous paragraph was also declaring her right of access to the communal pasturage of her natal community. Residence after marriage is often virilocal, but in such cases the wife does not lose her rights to herd her animals, and those of her husband and children, on the *bofedal* of her community of origin. Thus horizontal patterns of nomadic movements within Isluga territory are characterized by criss-crossing connections giving people access to pasture grounds associated with hamlets and more isolated homesteads to which they have inherited rights of usage. The maintenance of these pastures is the responsibility of the herders and herds. Llamas and alpacas void their dung on communal piles, which help to regenerate the surrounding pasture. For their part, herders have extended the areas of *bofedal* through the use of canals to irrigate lands adjacent to rivers and streams.

From the 1990s onward, there has been an increasing tendency for children to remain in formal education beyond the age of twelve. In the 1980s, young people wishing to receive secondary education became boarders at the school in Colchane, a settlement near the international frontier with Bolivia. Now families are more likely to set up house in the new town of Alto Hospicio in the desert near the coastal city of Iquique to enable their children to attend schools in an urban setting. This migration has removed parents and school-aged children from permanent residence in the highlands for a period of some years. It adds to the migration of male adults seeking waged labour that has been occurring increasingly since the 1980s. Adult women also seek work as commercial traders or in domestic service in cities.

This pattern of human dispersal from Isluga was partially (and paradoxically) fostered by the loan of money to people during the closing year of the Pinochet regime in 1989 to enable them to construct new mud brick houses with concrete floors and imported timber to support corrugated iron roofs. Since then, many of the scattered settlements in Isluga have acquired a more village-like appearance, following a campaign which, in some respects, mimics the resettlement of the Colonial Period. In its late twentieth-century manifestation, the increasing agglutination of communities involved residents in financial outlays which forced family members to migrate from Isluga in order to secure a cash income to repay the loans. Hence the attempt to coerce a nomadic people to form more stable residential units was undermined by the need to participate more fully in the cash economy. In this case the ideal of "civility", as expressed in the conversion of scattered settlements into communities with street-like arrangements, was associated with the neoliberal economics to which the Chilean regime aspired.

The social changes sketched here have occurred more recently than in communities elsewhere in the Andes (Valdes *et al.* 1983), or amongst parallel trajectories experienced by pastoralists in other parts of the world. On the basis of his fieldwork and a study of the historically documented sources, Ingold (1980) tracked an evolutionary trend from reindeer hunting to the herding and, subsequently, the ranching of reindeer. In some parts of the Andes, herders leave their llamas to follow their own cycle of movements unsupervised, while available family members herd their sheep. However, the ranching of camelids has not characterized developments in Isluga. Herding strategies continue to be based on the interaction between the social organization of the herders and that of their herd animals (the third principle listed above). In cases where family members are absent for long periods of time, herds are combined, for example those of adult brothers or of parents and adult children are joined together, and the family members club together to pay the wages of an assistant herder, often a migrant from Bolivia. The migrants are taught Isluga herding strategies by the remaining residents in the community in which they are employed.

In another part of the world, Dawn Chatty (1996, 192) observed that Harasiis pastoralist families in Oman have hired labourers from the Asian subcontinent to look after the larger livestock and to carry water to compensate for the absence of able-bodied men, who left the community in order to seek waged labour. However, this pattern, which parallels the strategies adopted by Isluga pastoralists, is taking place against a wider picture of expansion in the Arabian Peninsula of what Chatty (1996, 193) calls "large-scale ranching type systems". In her study of household subsistence and change in Oman, she pointed out that pastoralism is a way of life which is both admired and regarded with suspicion in the Middle East. At one extreme it is seen as "a throwback to an earlier stage of human development" and at the other "a unique and sophisticated adaptation of a harsh environment" (Chatty 1996, 1). As was the case with Isluga and its relationship with the Chilean government, Harasiis became the target of the desire of a Middle Eastern government to settle pastoralists because, according to Chatty (1996, 164) the governments of the region saw them "as signifiers of internal political problems".

Ethnographic Data in Archaeological Interpretations

Given the persistence of pastoralism as a way of life and its modification according to the social-political contexts in which pastoralists operate, both in the Andes and in other parts of the world, how might archaeologists make use of ethnographic material? Its persistence might seem to imply the continuation of a long tradition. However, probably it would be more accurate to speak of a continuity represented by different sorts of pastoralist economies that have arisen in the Andes and have undergone transformations through time. The ethnographic account presented here attempts to provide something of the texture of a herding way of life as practised from the mid-1980s to the early years of the twenty-first century. It indicates the role that herd animals have played in the spiritual/ritual life as well as in the subsistence activities of the community of Isluga. It also indicates some of the social changes that have been occurring during this period. Camelid herding has continued in the Andes, despite the vicissitudes occasioned by inter-related factors such as demographic change (of both the herds and the herders), political disapproval of nomadic ways of life and climatic uncertainties.

Other ethnographic accounts of pastoralist societies based on camelid herding in the South-Central Andes provide a wider context for the data from Isluga (see, for example, Boman 1908; Flores Ochoa 1979; Caro 1985; Göbel 1994; Bolton 2007). All these publications present material that is of use for archaeologists, but not simply as a quarry of easily available ethnographic analogies. There is a danger in cherry picking pieces of evidence that are convenient for the *ad hoc* interpretation of archaeological evidence without understanding fully the underlying rationale of the social principles operating in the context from which they have been extracted. What follows is a brief review of how ethnoarchaeological approaches have been tackled in the Andes and some suggestions for addressing research concerns particularly in contexts involving pastoralist and agro-pastoralist societies characterized by mobility or nomadism.

Ian Hodder (1982, 16) observed that analogy was an interpretive strategy commonly used by archaeologists. He identified its use in formal and relational terms. Archaeologists use formal analogy in cases where they consider two objects or situations with shared properties to have other similarities in common as well. This use of analogy occurs most frequently in archaeological interpretations. However, Hodder (1982, 16) pointed out that the association of the characteristics involved in the analogy "may be fortuitous or accidental". In contrast, in a relational analogy the investigator seeks to establish natural or cultural linkages between the different features involved, bearing in mind that these linkages are interdependent and not merely accidental. In his refinement in the use of analogy in archaeological interpretation, Hodder (1982, 23) stressed the importance of understanding the "links and contexts" of the aspects.

An example dating from the 1980s of the use of ethnographic analogy in an Andean context is provided by Helaine Silverman's (1986) study of the ceremonial centre of Cahuachi in southern Peru. In seeking to understand an apparent paradox (a centre with monumental architecture and an absence of evidence for permanent residence), Silverman and her colleague Miguel Pazos made several visits to a site of Catholic pilgrimage, the shrine of the Virgen del Rosario de Yauca in Ica. In her report, Silverman (1986, 471–474) offered a formal analogy in order to explain the lack of habitational refuse at Cahuachi by exploring the mechanisms by which open spaces (*plazas*) only used for specific events might be swept clean of rubbish through a combination of prevailing winds and human activity, the latter taking place immediately before the annual event at the Yauca shrine. Although not specifically mentioned by Kuznar (2001, 2–4), this pragmatic approach in the use of analogy would also correspond to situation which might be argued to display some aspects of historical continuity. Silverman did not develop the links and contexts in a relational analogy as advocated by Hodder, instead she extended her use of analogy by shifting to a different frame of reference. In effect, her discussion added a second analogy which relied on a differential extrapolation of data. Referring to the complexity of social organization among Australian aborigines, which is accompanied by a relatively modest material culture, Silverman (1986, 475) sought to unsettle "logical" expectations by arguing, conversely, that the monumental complexity of Cahuachi need not imply the existence of a centralized state or empire as envisaged by archaeologists who claimed Cahuachi as a putative capital.

A recent demonstration of the development of "relational analogy" combined with extrapolation has been provided by Jerry Moore (2005), but not in respect to pastoralism. His study of cultural landscapes in the Andes is based on a strategy of an extended ethnographic review of relevant literature from different parts of the world and extrapolation. This procedure enables him to consider social principles which are not normally visible in the archaeological record and how they might become accessible for analysis through the cultural acts performed by the members of the society in the past who were responsible for the residue which through these means become recognizable in that record.

Moore's strategy indicates the distance travelled by archaeologists since earlier discussions of ethnoarchaeological methods and applications. Stiles (1977) argued that "archaeological ethnography" involved the comparative study of variation in artefact form, of the spatial relationships visible in the patterning of cultural/economic activities and of disposal practices. He pointed out that such topics are of specific interest to archaeologists and are not necessarily recorded by ethnographers in their publications (Stiles 1977, 91). The basis for comparison between ethnoarchaeologically-observed and past practices rests on formal patterns of resemblance particularly as seen in artefacts and discarded items. Hence archaeologists have undertaken the collection of their own ethnographic data, but without the length of time spent in the field by social anthropologists, whose aim is to build up a holistic understanding of the underlying social principles that prevail in the community they are studying.

The work of Miller (1979) provided a milestone in camelid research in its detailed archaeofaunal study which he set against an examination of ethnographically and historically recorded methods of slaughtering camelids. However, following Chang and Koster's (1986) paper entitled "Beyond bones: Toward an ethnoarchaeology of pastoralism" attention turned to the investigation of settlement patterns. Both these approaches (faunal and settlement) depend on the recognition of formal patterns of resemblance between past and present, whether in the bones themselves or in settlement residue, as observed in the archaeological record. They assisted archaeologists to address the mechanisms by which the archaeological record is formed under specific conditions relevant to pastoralism. Yacobaccio *et al.* (1998, 18–19) provided a useful bibliography of ethnoarchaeological studies and they stressed the need to combine faunal studies with investigation into settlement patterns.

In an ethnoarchaeological study of goat herding in southern Peru, Kuznar (1995, 9) posed a question "whether or not these factors [*i.e.* the factors under which herding takes place] would have affected the archaeological record in the same way". If, by these factors, he referred to the inter-relationships between human herders, herd animals and pasture, his argument assumed that the differences between goats and camelids and their respective relationships with their owners are negligible. However, the feeding patterns of goats and camelids are different. Llamas are browsers and grazers and alpacas are grazers. Goats are browsers which degrade Andean shrubs under low goat-browsing pressures (Fuentes 1984, 47). Resource utilization is different for camelids and goats, too. While both supply meat and hides, the former provide their owners with fleece for spinning and can supply service as beasts of burden. In contrast, goats supply milk. Kuznar's argument also assumed unchanged ecological conditions. Research indicates that initial attempts to domesticate camelids in the South Central Andes probably occurred during moister climatic conditions (Dransart 2005, 286; Latorre *et al.* 2005, 86). Kuznar's strategy for using a subsistence strategy based on goat herding as an ethnographic analogy is only capable of leading to a generalized view of pastoralism in the past. The challenge to investigators is to conceive of pastoralism in its multiplicity.

Just as Ikeya (2006, 43) indicated that historically observed hunter-gatherer societies have multiple subsistence modes in which hunting and gathering activities are complemented by fishing, trading, herding and the cultivation of crops, pastoralism is likewise characterized by the resourcefulness of its practitioners. A study of $\delta^{13}C$ and $\delta^{14}N$ values in camelid bone remains from Conchopata in the highlands of Peru led Finucane *et al.* (2006, 1773) to identify two different camelid management practices during the period AD 550–1000. One was associated with maize cultivation in which camelids were fed maize stalks and hulks in contrast to the pasturing of camelids on grass and shrub lands.

The strategy I have previously explored (Dransart 2002), juxtaposed ethnographic evidence from contemporary Isluga with material evidence from archaeologically recuperated evidence of societies in the Atacama Desert. My objective was not to make direct analogies between Isluga practices and those of the distant past but, rather, to explore specific questions concerning the character of camelid pastoralism as it developed in different social contexts. It is therefore possible to detect differences in the relationships maintained between herders and their camelids in the past and to propose alternative forms of pastoral ways of life.

The social principles which I have identified as being relevant to an understanding of camelid pastoralism of the past in this paper are not amenable for examination using formal analogies. Yet it is these underlying social principles which ultimately make an impact on the biological processes of domestication, and which are of particular interest to zooarchaeologists. The use of ethnographic evidence in analogies can have the unfortunate effect of conveying the work of ethnographers to an archaeological readership of a timeless, unchanging present. This paper therefore invites archaeologists to consider carefully their use of ethnographic information by taking into account the changing social contexts in which the data were generated. By demonstrating how herding practices in a contemporary pastoralist society are changing, I hope to have presented the rationale behind those practices in a historically contingent context.

References

Apaza Suca, N., Komarek, K., Llanque Chana, D. and Ochoa Villanueva, V. (1984) *Diccionario Aymara-Castellano. Arunakan liwru Aymara-Kastillanu*. Puno, Proyecto Experimental de Educación Bilingüe.

Bolton, M. (2007) Counting llamas and accounting for people: livestock, land and citizens in southern Bolivia. *The Sociological Review* 55(1), 5–21.

Boman, E. (1908) *Antiquités de la région andine de la République Argentine et du Desert d'Atacama*. Paris, Imprimerie Nationale.

Caro, D. (1985) *Those who divide us. Resistance and change among pastoral ayllus in Ulla Ulla, Bolivia*. Ann Arbor, Michigan, University Microfilms International.

Chang, C. and Koster, H. A. (1986) Beyond bones: toward an ethnoarchaeology of pastoralism. In M. B. Schiffer (ed.) *Advances in Archaeological Method and Theory* 9, 97–148. Orlando, Academic Press.

Chatty, D. (1996) *Mobile pastoralists: development planning and social change in Oman*. New York, Columbia University Press.

Donkin, R. A. (1979) *Agricultural terracing in the aboriginal New World*, Viking Fund Publications in Anthropology No 56. Tucson, Wenner-Gren Foundations for Anthropological Research.

Dransart, P. (2002) *Earth, water, fleece and fabric. An ethnography and archaeology of Andean camelid herding*. London and New York, Routledge.

Dransart, P. (2005) Living with llamas at 23° S. In A. Smith and P. Hesse (eds.) *23° South: Archaeology and Environmental History of the Southern Deserts*, 281–291. Canberra, National Museum of Australia.

Dransart, P. (forthcoming) Animals and their possessions: properties of herd animals in the Andes and Europe. In M. Bolton and C. Degnen (eds.) *Animals and Science: from colonial encounters to the Biotech industry*. Newcastle upon Tyne, Cambridge Scholars Publishing.

Finucane, B., Maita Agurto, P. and Isbell, W. H. (2006) Human and animal diet at Conchopata, Peru: stable isotope evidence for maize agriculture and animal management practices during the Middle Horizon. *Journal of Archaeological Science* 33, 1766–1776.

Flores Ochoa, J. (1968) *Los pastores de Paratía. Una introducción a su estudio*. Serie Antropología Social No 10. Mexico, Instituto Indigenista Interamericana.

Flores Ochoa, J. (1979) *Pastoralists of the Andes. The Alpaca Herders of Paratía*, translated by R. Bolton. Philadelphia, Institute for the Study of Human Issues.

Forde, C. D. (1934) *Habitat, economy and society*. London, Methuen.

Fraser, V. (1990) *The architecture of conquest: building in the Viceroyalty of Peru, 1535–1635*. Cambridge, Cambridge University Press.

Fuentes, E. R. (1984) Human impacts and ecosystem resilience in the southern Andes. *Mountain Research and Development* 4(1), 45–49.

Göbel, B. (1994) El manejo del riesgo en la economía pastoril de Susques. In D. E. Elkin, C. M. Madero, G. I. Mengoni, D. E. Olivera, M. C. Reigadas and H. D. Yacobaccio (eds.) *Zooarqueología de Camélidos I*, 43–56. Buenos Aires, Grupo Zooarqueología de Camélidos.

Hodder, I. (1982) *The present past: an introduction to anthropology for archaeologists*. London, Batsford.

Ikeya, K. (2006) Mobility and territorialilty among hunting-farming-trading societies: the case study of bear hunting in mountain environments of Northeastern Japan. In C Grier, J. Kim and J. Uchiyama (eds.) *Beyond affluent foragers: rethinking hunter-gatherer complexity*, 34–44. Oxford, Oxbow Books.

Ingold, T. (1980) *Hunters, pastoralists and ranchers*. Cambridge, Cambridge University Press.

Ingold, T. (1986) *The appropriation of nature: essays on human ecology and social relations*. Manchester, Manchester University Press.

Kuznar, L. A. (1995) *Awatimarka: the ethnoarchaeology of an Andean herding community*. Fort Worth, Harcourt Brace College Publishers.

Kuznar, L. A. (2001) Introduction to Andean ethnoarchaeology. In L.A. Kuznar (ed.) *Ethnoarchaeology of Andean South America: contributions to archaeological method and theory*, 1–18. Ann Arbor, International Monographs in Prehistory.

Latorre, C., Betancourt, J. L., Rech, J.A., Quade, J., Holmgren, C., Placzek, C., Maldonado, A. L. C., Vuille, M. and Rylander, K. (2005) Late Quaternary history of the Atacama Dessert. In A. Smith and P. Hesse (eds.) *23° South: Archaeology and Environmental History of the Southern Deserts*, 73–90. Canberra, National Museum of Australia.

MacCormack, S. (1991) *Religion in the Andes: Vision and Imagination in Early Colonial Peru*. Princeton, Princeton University Press.

Mamani Mamani, M. (2002) *Diccionario práctico bilingüe Aymara-Castellano. Zona Norte de Chile. Suman chuymamp parlt'asiñi*. Antofagasta, EMELNOR NORprint.

Martínez, G. (1976) El sistema de los uywiris en Isluga. *Anales de la Universidad del Norte (Homenaje al Dr Gustavo Le Paige S.J.)* 10, 255–327.

Mengoni Goñalons, G. L. and Yacobaccio, H. D. (2006) The domestication of South American camelids: a view from the South-Central Andes. In M. A. Zeder (ed.) *Documenting Domestication: New Genetic and Archaeological Paradigms*, 228–244. Berkely, University of California Press.

Miller, G. R. (1979) *An introduction to the ethnoarchaeology of the Andean camelids*. Ann Arbor, Michigan, U.M.I.

Milner, N. and Miracle, P. (2002) Introduction: patterning data and consuming theory. In P. Miracle and N. Milner (eds.) *Consuming passions and patterns of consumption*, 1–5. Cambridge, McDonald Institute Monographs.

Monast, J. E. (1972) *Los indios aimaraes. ¿Evangelizados o solamente bautizados?* Buenos Aires, Cuadernos Latinoamericanos Ediciones Carlos Lohlé.

Moore, J. D. (2005) *Cultural landscapes in the ancient Andes: archaeologies of place*. Gainesville, University Press of Florida.

Ortega Perrier, M. (1998) 'By reason or by force': Islugueño identity and Chilean nationalism. Unpublished PhD thesis. Department of Social Anthropology, University of Cambridge.

O'Sullivan, D. (2001) Space, silence and shortage on Lindisfarne. The archaeology of asceticism. In H. Hamerow and A. MacGregor (eds.) *Image and power in the archaeology of early medieval Britain. Essays in honour of Rosemary Cramp*, 33–52. Oxford, Oxbow.

Silverman, H. (1986) La investigación arqueológica y el uso de la analogía etnográfica: el caso de las plazas y espacios abiertos de Cahuachi. *Revista Andina* Año 4(2), 465–478.

Stiles, D. (1977) Ethnoarchaeology: a discussion of methods and applications, *Man* (N.S.) 12, 87–103.

Valdes, X., Montecino, S., Leon. K. de, and Mack, M. (1983) *Historias testimoniales de mujeres del campo*. Santiago, PEMCI.

Wheeler, J. C., Chikhi, L. and Bruford, M. W. (2006) Genetic analysis of the origins of domestic South American camelids. In M. A. Zeder, D. G. Bradley, E. Emshwiller and B. D. Smith (eds.) *Documenting domestication: new genetic and archaeological paradigms*, 329–341. Berkeley, University of California Press.

Yacobaccio, H. D., Madero, C. M. and Malmierca, M. P. (1998) *Etnoarqueología de pastores surandinos*. Buenos Aires, Grupo Zooarqueología de Camélidos.

14. Incidence and causes of calf mortality in Maasai herds: Implications for zooarchaeological interpretation

Kathleen Ryan and Paul Nkuo Kunoni

Maasai are recognized as expert husbandmen; every effort is made to support their livestock. Despite these efforts, many calves still succumb. Deaths in the first year of life have been recorded as low as 5% (in good years) and as high as 100% (in years of prolonged drought or epidemic disease). Few are deliberately culled. This paper focuses on age at death, cause of death, and attempts at prevention or treatment. Data were gathered from three early life stages (birth to 3 months, 3 to 6 months, and 6 to 12 months) and correlated to a bimodal rainfall pattern of "long" and "short" rains. Environmental assaults such as prolonged drought, flash floods, heat or cold were also documented. Numbers of calf deaths per annum were compared to mortality in the adult herd. To provide a comparative perspective data were collected from other geographic areas including other areas of Africa, Canada, and the United States. In each case, figures were recorded separately for deliberate culling (if any) and death from disease or disaster. These data have implications for interpretation of archaeological faunal assemblages created under similar environmental conditions in the past.

Keywords: Maasai, calf mortality, zooarchaeology, Africa, Canada, United States

Introduction

Zooarchaeologists frequently turn to ethnographic or historical records to inform their models of the organization and economics of pastoral societies. In this paper we focus on cattle ecology and examine how cattle cope with a great diversity of environments in many parts of the world. Specifically, we attempt to document the incidence and possible causes of calf mortality. Our interest in this early life stage was piqued by the questions raised by Mulville *et al.* (2005) in their study of calf mortality from prehistoric contexts in the Northern and Western Isles of Scotland. Why did so many neonates die and were their deaths due to natural causes or to human intervention? And, if it was the latter, was it due to poor management or deliberate culling? The presence of large numbers of young calf remains in archaeological contexts has generated a lively debate surrounding their interpretation. Were calves deliberately culled as part of a dairying strategy or did they die of natural causes such as disease or harsh environment (Bogucki 1986; Crabtree 1986; 1990; 2003; Halstead 1998; Legge 1981a; 1981b; 1992; 2005; McCormick 1998; Mulville *et al.* 2005)? It is not our intention to favour one over another of these options; rather we present a series of modern case studies from a diversity of contexts where calf mortality was moderate to high and where the cause of death was known.

In our study cattle remain the constant around which other variables (such as husbandry practices, environment and time period) revolve. We include ethnographic data from East Africa; recent veterinary records from North Africa, Canada, and the United States; ethnohistorical data from early medieval Ireland and early modern Scotland; and zooarchaeological data from East Africa, to develop expectations for zooarchaeological remains in general and calf remains in particular, in order to shed light on cattle husbandry practice in prehistoric contexts. Cattle are the common thread in all of these studies.

Research Design

Our plan was to track calf mortality in several Maasai herds over a four year period, recording age at death, cause of death, and attempts at prevention or treatment. Fluctuations in weather patterns such as excessive heat or cold, prolonged drought beyond the normal dry seasons, or flash floods were also recorded. These data were compared with mortality data from dairy herds in other areas of Africa and from Canada and the United States, from historical

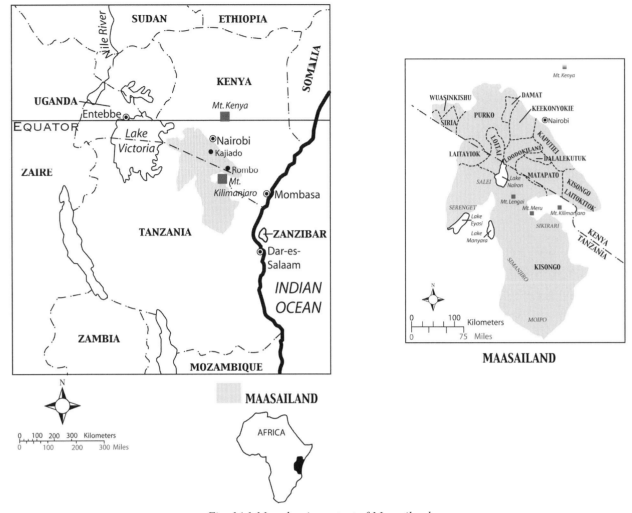

Fig. 14.1 Map showing extent of Maasailand

commentaries relating to cattle husbandry from the 17th and 18th centuries in Scotland, and from early Irish Law tracts (7th to 8th century AD). In addition, our results were compared to mortality rates from a well preserved faunal assemblage from an archaeological site in Kenya's Maasailand where there was relatively good preservation of neonates (Marshall 1990).

Environmental Context of Maasailand

Although straddling the equator, environment and climate vary considerably throughout East Africa, ranging from the hot and semi-arid areas of northern Kenya to the cool conditions of the highlands. Several environmental/climatic zones can be distinguished on the basis of vegetation and rainfall (Hamilton 1982; Lind and Morrison 1974). Cultivation is dependent not only on rainfall but on *predictable* rainfall. Mobile pastoralism, on the other hand, can cope in areas of lower and less predictable rainfall by virtue of being mobile – stock can be moved to within reach of locally remaining water sources when dry conditions are prolonged, and distributed widely in the aftermath of the rains when grasses are both abundant and nutritious.

Most of the mobile pastoral groups of East Africa inhabit arid or semi-arid landscapes. Galaty (1994, 200) estimated the pastoral population of Kenya at 3.5 million. Maasai cattle herders today inhabit areas of wooded grassland and bushland thicket with low and unpredictable rainfall in southern Kenya and northern Tanzania (Hamilton 1982). Our study sites are in Kenya's Kajiado District south of Nairobi as far as the Tanzanian border (Fig. 14.1), which covers 21,105km^2 with a human population reported in 1979 by Meadows and White (1979, 2) of approximately 107,000 of whom 70,000 were Maasai.

Methodology

Our previous studies of Maasai cattle mortality patterns focused mainly on adult animals (Ryan *et al.* 2000). Data were gleaned from detailed cattle genealogies stretching back 130 years. Female cattle were named for the lineage into which they were born. Heifers did not get named officially until they were at an age to breed or when they dropped their first calf and were therefore likely to survive and pass on the name of the lineage to their progeny. Neither bulls nor steers carried the lineage name although their

Year	Long rains: March to May	Short rains: between Oct and Dec
2003	below average	below average
2004	below average	below average
2005	average	average
2006	average	average

Table 14.1 Duration and intensity of rains 2003–2007

lineages were known and remembered. Although deaths of calves were mentioned by our Maasai consultants, they did not feature as a major part of the study of individual cattle lineages.

The traditional Maasai management pattern, under unstressed conditions, favoured the retention of females into old age, as long as they continued to reproduce. Most males were castrated at a young age but were not slaughtered until they were in their prime, around 4 to 5 years of age. A few males were kept intact for breeding purposes and were normally slaughtered around 8 to 10 years of age when they were replaced by younger bulls.

For the present study, data were gathered from three early life stages (birth to 3 months, 3 to 6 months, and 6 to 12 months) and correlated to a bimodal rainfall pattern of "long" and "short" rains. Long rains are expected to occur in March to May, short rains any time between October and January, although there is quite a lot of local variation in the predictability and duration of rains in any year. Patterns appear to be even more unpredictable in recent years as pastoralists and agriculturalists alike try to cope with prolonged drought or flash floods. Many blame these vicissitudes on climate change brought on by global warming. For this study therefore, environmental assaults such as prolonged drought, flash floods, heat or cold were also documented. Where possible, numbers of calf deaths per annum were compared to mortality in the adult herd. In each case, data on death from disease or disaster were recorded separately from deliberate culling (if any).

Between March 2003 and February 2005 rains were well "below average," a euphemism for almost non-existent. Rains in 2005 were average with long rains coming in February, earlier than normal, and in 2006 rains were above average (Table 14.1). In a good year cattle will be at their nutritional best in June/July, will conceive, and subsequently drop calves in March/April. If they conceive after the short rains, calves can be born any time from July through October depending on the time of onset and the duration of the rains. In Maasailand cows' milk yield generally tapers off after the third month of a new pregnancy.

Datasets

Data were gathered from three areas of Maasailand (Fig. 14.1) occupied by four separate subsections of Maasai: the *Kaputiei* just south of Nairobi; the *Dalalekutuk*, near Kajiado town; the *Kisongo (Laitokitok)*, in and around Rombo and the foothills of Kilimanjaro; and the *Matapato* near Namanga. Individual consultants were drawn from the families of Lengete Oldukunyi *(Kaputiei)*, in Upper Kaputiei; Keeja ole Leeyio *(Kaputiei)*, in Kitengela; Kasimiro ole Kaaka *(Dalalekutuk)*, in Isajiloni; Parmitoro ole Koringo (*Kisongo Laitokitok*), in Enkusero and Ol Girra, near Rombo; and the extended Olosikeri family *(Matapato)*, in Metto near Namanga.

Oldukunyi Family Ranch

The Oldukunyi family ranch is relatively large, located about 50km south of Nairobi in the Upper Kaputiei area. Calves are normally born in this area between January and March or between October and December. One hundred and seventy calves were born at the Oldukunyi ranch between 2003 and 2006, none died in the birth–3 month period; 14 died in the 3–6 month period; and 19 in the 6–12 month period. At this ranch, which has access to veterinary care, the main killer is a tick borne disease, East Coast Fever (ECF); the main disaster factor is drought leading to starvation. Of calves that died in 2003, two (aged 3 to 6 months) died of ECF and eight (aged 6 to 12 months) due to drought. None died in the birth–3 month period. In 2004, three died of ECF, five due to drought (all 3 to 6 months old) and two due to coccidiosis. In 2005, of 15 calf deaths two were from ECF and 13 were due to drought. Of the 35 calves born between January and June 2006 (the rains finally came in February) two died of ECF and three of coccidiosis, a parasitic disease which causes severe diarrhea. According to West (1985), young animals are usually infected by this parasite at pasture between the time they first go out to graze and two years of age. In January to June 2007, another four calves were born, one died of pneumonia. Further data on this herd were gathered in 2007; see Table 14.2.

The adult herd reflected the same general mortality patterns: death from ECF or due to drought. In 2004, three bulls and one older cow died of ECF and three older cows and 10 heifers died due to drought. In 2005, five older cows, three heifers, and one bull all died due to drought.

Ole Leeyia Family Ranch

Data gathered from the ole Leeyia family ranch in Kitengela are presented in Table 14.3. Of the relatively large cohort of 56 calves born January to June 2004, 30 died due to drought. Later that same year another 22 were born and six died, one of ECF and five due to drought. Birth rates were low (below 20) in January to June 2005 as the drought

YEAR	BORN			DIED			CAUSES OF DEATH
	TOTAL	MALES	FEMALES	TOTAL	MALES	FEMALES	
2004 JAN – JUNE	22	12	10	9	3	6	2 E.C.F. 2 Cocidiosis 5 Drought
JULY – DEC	10	6	4	1	1	-	E.C.F.
2005 JAN – JUNE	32	10	22	5	4	1	3 Drought 2 E.C.F. 2 Aborted (not counted as calf)
JULY – DEC	35	9	26	10	6	4	Drought
2006 JAN – JUNE	35	20	15	5	3	2	2 E.C.F. 3 Coccidiosis
JULY – DEC	10	3	7	-	-	-	1 Aborted (not counted)
2007 JAN – JUNE	4	2	2	1	1	-	Pneumonia The dam died of Ephemeral fever (3 days sickness)

Table 14.2 Calf mortality data for Oldukunyi family ranch

YEAR	BORN			DIED			CAUSES OF DEATH
	TOTAL	MALES	FEMALES	TOTAL	MALES	FEMALES	
2004 JAN – JUNE	56	20	36	30	17	13	Drought
JULY – DEC	22	7	15	6	4	2	1 E.C.F. 5 Drought
2005 JAN – JUNE	18	9	9	3	2	1	Drought
JULY – DEC	25	8	17	2	2	-	2 ECF
2006 JAN – JUNE	15	10	5	3	2	1	2 ECF 1 Black Quarter
JULY – DEC	10	4	6	1	1	-	Pneumonia
2007 JAN – JUNE	8	2	6	-	-	-	-

Table 14.3 Mortality data for ole Leeyio family ranch

continued. Numbers picked up in July to December 2005 with only two deaths from ECF and none from drought. Birth rates were again low from 2006 to 2007 but mortality was also reduced. Although we have no data for mortality of the adult herd, the reduced birth rates suggest that mortality of breeding cows was high.

Ole Kaaka Family Boma

The ole Kaaka family holding is located in Isajiloni, about 60km south of Nairobi, close to Kajiado town. This is a traditional Maasai settlement (*boma*). A roughly circular enclosure (Fig. 14.2) with a central cattle corral is shared by all families that occupy the boma, and separate subsidiary

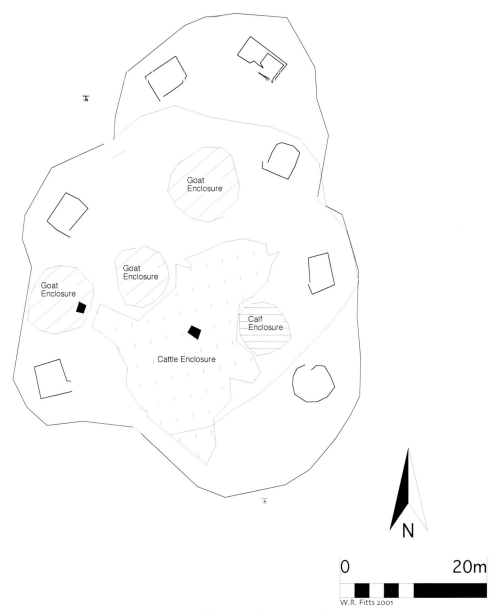

Fig. 14.2 Schematic of Maasai settlement

sheep/goat enclosures exist for each family. Family houses are built around the periphery and the whole settlement is enclosed in a thorn fence. Individual gates allow access to each family's houses and to the central corral.

Calves at Isajiloni are normally born between March and May (during or after the long rains) and October to January (which may or may not coincide with the short rains). Between 2003 and 2006, seven calves died before 3 months of age, but this figure includes late abortion or stillborn calves; three calves died in the 3–6 month period; and two between 6 and 12 months. We were told that the circumstances surrounding death included abortion, death of dam, decreased immunity to disease, harsh environment, lack of food, lack of veterinary care, predators, parasites, and direct killing of a calf in favour of the mother's survival although this last is *extremely* rare. Diseases prevalent in the area include tick borne diseases, such as ECF; anaplasmosis, an infectious disease of cattle caused by a parasite; pneumonia from cold temperatures; foot and mouth disease; blackquarter, an acute infectious disease usually found in young calves 3 to 6 months old; and lumpy skin disease caused by the Neethling pox virus.

Ole Koringo Family Bomas

At Enkusero, in the foothills of Kilimanjaro, and Ol Girra, near Rombo, are two traditional bomas, roughly 40km apart, occupied by the ole Koringo family. In this area calves are normally born from early May through June, at the end or just after the long rains. In January 2006 Enkusera had only nine cattle remaining of which six were female (four adult cows and two heifers) and three steers. During the period 2003–2006, six calves were born, two died: one male died between birth and three months

YEAR	BORN			DIED			CAUSES OF DEATH
	TOTAL	MALES	FEMALES	TOTAL	MALES	FEMALES	
2004 JAN – JUNE	9	6	3	6	3	3	3 males eaten by lions 1 female fell into a ditch 2 Black Quarter
JULY – DEC	8	3	5	1	1	-	Snake bite
2005 JAN – JUNE	7	4	3	-	-	-	-
JULY – DEC	4	2	2	-	-	-	-
2006 JAN – JUNE	6	3	3	-	-	-	-
JULY – DEC	5	3	2	1	1	-	Black Quarter
2007 JAN – JUNE	3	1	2	-	-	-	-

Table 14.4 Calf mortality data for ole Koringo family holdings

YEAR	BORN			DIED			CAUSES OF DEATH
	TOTAL	MALES	FEMALES	TOTAL	MALES	FEMALES	
2004 JAN – JUNE	84	44	40	11	6	5	All by East Coast fever
JULY – DEC	22	11	11	2	1	1	Drought
2005 JAN – JUNE	104	60	44	75	31	44	Drought
JULY – DEC	20	9	11	9	3	6	Drought
2006 JAN – JUNE	44	20	24	-	-	-	-
JULY – DEC	20	14	6	-	-	-	-
2007 JAN – JUNE	42	14	28	2	2	-	Diarrhea / worms

Table 14.5 Calf mortality data for Olosikeri family holdings

from trypanosomiasis in 2005; one female died at 2 1/2 weeks from foot and mouth disease in 2006; three males remained in the herd.

In January 2006 there were 30 adult cattle in the Ol Girra herd: 25 females, two bulls, and three steers. Data gathered in 2007 to cover 2004 to 2007 are shown in Table 14.4. The highest mortality occurred in 2004 when six out of nine live-born calves died: three male calves were eaten by lions, one female calf fell into a ditch, and two females died from black quarter. All were older than 3 months of age. Overall reproduction was low but it should be noted that this herd lost only two calves between January 2005 and June 2007.

Olosikeri Family Ranch

The Olosikeri family holdings comprise a large extended family operation under one management, located about 120km south of Nairobi, at Metto near Namanga close to the border with Tanzania. Mortality data for 2004–2007 are presented in Table 14.5. The highest mortality occurred in 2005 (January to June) when 75 out of 104 calves died as a result of drought. No mortality occurred in 2006 after the rains came, but it should be noted that the number of calves born in 2006 was only 64 compared to 106 in 2004 and 124 in 2005, suggesting that the drought had taken a serious toll of the adult cows in the herd. In January to June 2007, another 42 calves were born. Two male, 3 to 6 months of age, died of diarrhea or worms.

We have no precise data for calf age at death in the ole Leeyio and Olosikiri family herds but we were assured that very few calves died in the birth–3 month period in either herd. The majority died after 3 months of age.

Comparative Herds

For comparative purposes the Maasai data were compared with those from (i) a non-Maasai area in Central Kenya where agriculturalists kept a few dairy cows; (ii) a very large dairy operation in Libya; (iii) a smaller dairy operation in Ontario, Canada, and; (iv) a selection of 30 dairy farms in southeast Minnesota. At the Central Kenyan farm (i), age at mortality followed most closely that observed in the Maasai herds, but at most of these sites (ii–iv), calf mortality was highest in the birth–3 month period and lessened as they got older, a pattern which is the opposite of the majority of Maasai cases where mortality normally increased with age, usually after weaning, when young stock became dependent on available vegetation.

Central Kenya

In the Central Kenyan case (Gitau *et al.* 1994), farmers reared their calves as replacement herd animals and therefore there was every incentive to keep them healthy for that reason. Despite that, morbidity was high. In most cases, calves suffering from a variety of ailments at a young age were those that succumbed before the end of their first year. Patterns of morbidity and mortality were tracked for 78 of 90 randomly selected dairy farms in Kiambu District. Overall, 201 calves (104 males and 97 females) were observed on at least one visit. Calf mortality in the first year of life was 21.6%. The main cause of death was enteritis. Calf morbidity and subsequent mortality increased with age, particularly after weaning when calves were taken out to pasture.

Libya

In the Libyan case (Gusbi and Hird 1982/1983), not a great deal of effort was put into the survival of the calves. Calf mortality was studied on five Libyan dairy farms during a 5-year period (1976 through 1980). During that time 7325 calves were born, 3571 (49%) males and 3754 (51%) females. Of those, 1196 (16.3%) calves died between birth and 30 days (580 [48.5%] female, 616 [51.5%] male). Between 31 and 90 days 181 (7%) died (approx. 50% female, 50% male). Of the 1196 calves dying before 31 days of age, 890 (75%) died in the first week of life, 238 (20%) died between 8 and 14 days; 66 (0.55%) died between 15 and 20 days; and 2 (0.002%) died at 22 to 30 days. Overall, 18.8% of calves died in their first 3 months of life and, unlike in the Maasai examples, mortality decreased with increasing age.

Summarizing the causes of mortality overall, and particularly in the early life stage, the authors of the Libya study investigated differences in management practice. Station 5 had the lowest calf mortality rate. It was a newly established station, all cows were born in Libya, and its manager was more experienced than managers at some of the other stations. At Station 3 the calf mortality was higher overall than at Station 5 but the mortality of heifer calves was the lowest of all stations, while the male calf mortality was approximately the same as the average at other stations. The differences may be explained by the calf caretakers who were recruited from local farmers with experience on their own farms. In addition, their background and experience may have biased them to value female calves more than male. Station 4 had the highest number of births and the lowest survival rate for both female and male calves. Overcrowding and the inexperience of the caretakers (foreign workers who had no previous experience of livestock care) may have contributed to high mortality at this station. None of the calves on any of the stations were raised as replacement herd animals so there was less incentive to expend effort on their care.

Ontario

A study of a dairy operation in Ontario, Canada (Waltner-Toews *et al.* 1986a; 1986b; 1986c) focused on heifer calf management, morbidity, and mortality in Holstein herds between October 1980 and July 1983. Data were collected at farm level (to include all farms) and at the individual calf level (1968 calves from 35 farms). Farm production ranged from 23 to 154 calvings per year. Of the 1968 live-born calves 4% died before the age of weaning, *i.e.* approximately three months. This is a much lower mortality rate than recorded in Libya at 18.8%. No data were available for the older age groups as the study focused specifically on mortality of calves from birth to three months of age.

Associations between calf management and morbidity, enteritis and pneumonia in particular, were also examined. Enteritis and pneumonia were significantly associated with each other at both the farm and the calf level. No significant associations were found between individual calf management and the odds of being treated for enteritis, whereas many factors including sire use, method of first colostrum feeding, administration of anti-enteritis vaccine in the dam and preventive antimicrobials to the calf significantly affected the odds of a calf contracting pneumonia (Waltner-Toews *et al.* 1986c). The importance of colostrum to the young calf is discussed below.

Minnesota

A study of 845 heifer calves born during 1991 on 30 Holstein dairy farms in southeast Minnesota focused on calf morbidity and mortality from birth to 16 weeks of age (Sivula *et al.* 1996). Sixty-four calves died during the 16 months of the study. Of these 28 (44%) died as a result of enteritis and 19 (30%) of pneumonia. The risk of enteritis was highest in the first three weeks of life, and pneumonia at highest risk at 10 weeks. Risk of death overall was highest at two weeks of age. As with the previous study data are not available for the older age groups.

Ethnohistorical Data

Historical commentaries relating to Scotland and the Scottish Isles cited by McCormick (1998) provide evidence closer geographically and in environment to the sites discussed in Mulville *et al.* (2005). Accounts from the 17th century describe farmers' worry about the timing of births. An early birth, before the beginning of March, posed a problem for both the cow and its calf since spring grasses could not be expected before April; or the cow may have been in such a poor nutritional condition as to make it unable to give sufficient milk and the calf would be likely to die of starvation. Either of these scenarios could account for the high rate of mortality in the first month of life. Some, however, may have been culled to remove them from their dams after lactation had been initiated so that the dams would survive and continue to give milk in the absence of the calf, although our experience in Maasailand (and in other traditional herding systems) suggests that the dam is more likely to continue lactating if her calf is present (Ryan 2005). Late calving also posed a problem, as the calf would have difficulty picking up enough condition to survive its first winter. Deaths of late born calves between the ages of six and nine months could be expected any time from November through April. Some late born calves may therefore have been deliberately culled at the beginning of winter. Livestock mortality in Scotland as a whole was estimated at one in five during the winter in the 18th century (McCormick 1998, 50).

McCormick's reference to the 19th century Hebridean practice of raising one calf with the milk of two cows, implying that the excess calf was killed at the beginning of the lactation (McCormick 1998 citing McKay 1980, 65) could be a deliberate killing or a natural death. We noted a version of this practice in Maasailand where a calf was set to suckle from two cows when one cow had lost its own biological calf. This is intended to promote and prolong lactation in the cow whose calf has died. An interesting comment relating to this practice appears in an Old Irish law tract adjudicating the rights of ownership of a calf with two mothers:

A cow that bears a dead calf, which is put onto another calf, (*i.e.* it belongs to the cow which bears [it] only); it belongs to the mother alone, for that is not done for the sake of the calf but for the sake of the milk of the cow (*i.e.* which bore the dead calf). (Egerton 88 manuscript printed in *Corpus iuris hibernici* [CIH] i–vi [Binchy 1978] quoted in Kelly 1997.)

Many of the Old Irish law texts provide information on livestock and farming and may be of some interest in this discussion. The main Old Irish text on base clientship, *Cáin Aicillne*, distinguishes four categories of castrated male cattle that can be given from a client to his lord for food-rent. The first are calves between birth and 6 months and the second calves between 6 months and one year. These healthy calves were deliberately slaughtered to make food-rent payments from a client to his lord (Kelly 1997 citing *CIH* ii 483.28). Calves were classified by name, age and value. Thus a young male calf (*lóeg*) born in the spring (birth to 6 months) is the smallest deemed suitable for payment of food-rent which a client gives to his lord (Kelly 1997, 59).

The calf is only acceptable if it has a minimum girth of eight fists (probably 32 inches), and is sufficiently plump that its haunches cover its kidneys, except for the space of three fingers. It is expected to have grazed on grass with the milch cows, and its castration wound must have healed (*slán ó chull*). It must not have died of sickness or disease, but have been slaughtered by its owner. It is suitable for cooking in summer.

Under this system, one would expect to see calves as young as 3 to 6 months being slaughtered for food in the summer with a further cohort of 6 months or older at the beginning of winter. Keeping calves longer than November would not have increased their condition as the fodder became increasingly sparse. The use of male calves in these food-rent transactions suggests that it was a means of culling young males surplus to the need for traction animals or herd growth, especially in a society that according to all historical accounts was heavily dependent on dairy production.

Zooarchaeological Data

To compare our Maasai data to an archaeological context created under somewhat similar conditions, we chose the site of Ngamuriak, an open air site located in the Lemek Valley, in the Loita-Mara region of Kenya, excavated by Robertshaw and Marshall (Robertshaw 1990); zooarchaeological analysis and commentary carried out by Marshall (1990). The site produced a relatively large number of cattle remains, (MNI=26; NISP=2228). Based on dental samples, as percentages of MNI (N=26), Ngamuriak cattle age distribution is estimated as follows: 11% of calves died in the first six months; 23% of juveniles died between 6 and 18 months; 2% of juveniles between 18–30 months; 18% of young adults between 18–48 months; 40% of adults between 4 to 9 years; 3% aged between 9 and 14 years (Marshall 1990, Fig. 10.2, 215). As we have demonstrated from our study of Maasai calf mortality, few died in the birth–3 month period, with a slight increase in the 3–6 month period, and a further increase in the 6–12 month period. From our previous genealogical studies (Ryan *et al.* 2000), we know that steers were kept to 4 or 5 years of age before slaughter, bulls tended to be culled at 8 to 10 years, and cows were allowed to live out their natural life, 15 years and up, especially if they were still breeding. The comparison would confirm Marshall's suggestion that Ngamuriak cattle herders were managing their herds in ways very similar to modern Maasai (Ryan *et al.* 2000).

Fig. 14.3 Newborn calf being presented for inspection

Summary of Mortality Patterns

In summary, our recent studies in Maasailand revealed a pattern of mortality where calves tended to survive the first three months of life but were at increasing risk as they aged. Usually the risk increased after weaning, any time from three to six months, when they became dependent on foraging on the available vegetation. Our data suggest that over the four years of our study drought leading to morbidity or starvation was the main killer. Additional risk lay in the vegetation itself where parasites lurked or from diseases transmitted by vectors in the air or by contact with other animals.

This pattern is not repeated in any of our other modern studies. Even the Central Kenyan study which also had a higher risk of mortality in the older age group is difficult to compare to the Maasai data because farmers had access to grains to supplement feeding and no calves died of starvation due to drought. In all of the other case studies (Libya, United States or Canada), mortality was either very high or moderately high in the first three months of life. In fact, the first two weeks of a calf's life appeared to present the greatest risk of death, usually related to enteritis, followed by another peak at around 10 weeks when many succumbed to pneumonia. There is no appearance of neglect, except in the Libyan case where the hiring of unskilled workers clearly affected calf mortality.

Fig. 14.4 Calf suckling soon after birth

The data gathered from 17th and 18th century sources relating to Scotland and the Scottish Isles estimate an overall mortality of one in five during the winter (McCormick 1998, 50). It is clear, however, that ensuring the survival of young animals was fraught with potential disaster. Difficult decisions would have to be made, many of them driven by the physical environment, exacerbated by disease. It would be difficult to determine if calves were deliberately culled to create the ideal herd balance for a dairy economy, and thus more milk for humans, or if calves were slaughtered when the dam died or was at risk of death.

As for the early Irish cases of deliberate culling, described in the law texts, one must bear in mind that the laws are "proscriptive" (or ideal), not descriptive of what was actually happening. They do, however, lay out a clear sequence that would indeed follow a pattern of deliberate culling suitable for an emphasis on dairy production.

Fig. 14.5 Day old calf being comforted by elder

Why do Calves Survive?

We have seen from the above studies why many calves die. In Maasailand it is certainly not from neglect or poor management, rather it is more likely to be from environmental or physiological assaults such as drought or disease. As we have already mentioned, common diseases include East Coast Fever, anthrax, rinderpest, foot-and-mouth, trypanosomiasis, anaplasmosis, coccidiosis, and blackquarter, among others. We should look also at why so

Fig. 14.6 Young calves form age cohorts and graze close to the settlement

Fig. 14.7 At weaning time calves are encouraged to eat vegetation

Fig. 14.8 Older calves are grouped to graze away from the settlement

Fig. 14.9 As each calf reaches an appropriate age it is introduced to the group

many calves in Maasailand *survive* despite these potential assaults.

Husbandry

The Maasai recognize that it is essential that a cow bonds with her newborn immediately after the birth (Ryan 2005). In Fig. 14.3, the owner is presenting a calf to its mother for inspection and in Fig. 14.4, only minutes later the calf is already suckling. It is essential that the calf gets the colostrum, the milk secreted by the udder immediately after parturition and for the following three to four days. It contains 20% or more protein and a little more fat than normal milk. According to West (1985):

> It is normally rich in vitamins A and D provided the dam has not been deprived of these in her food. It acts as a natural purgative for the young animal, clearing from its intestines the accumulated faecal matter known as 'meconium', which is often of a dry putty-like nature. Of much greater importance, it is through the medium of the colostrum that the young animal obtains its first supply of antibodies which protect it against various bacteria and viruses.

However, these antibodies are difficult for the calf to absorb, and may cause enteritis, but West reports that it has been demonstrated that "the physical presence of the dam with the calf, in some unknown way, facilitated the absorption of immunoglobulin."

Maasai take great care of the calf during the first few months of life, especially in the first month. The calf pictured in Fig. 14.5 is only one day old. He is separated from his mother during the day when she goes out to graze and is suckled by her morning and evening during milking. Two teats are normally reserved for the calf. If a calf dies, its skin is put over another calf, sometimes an orphan but sometimes over a calf that has a mother. In the latter case the calf benefits from the milk of two mothers and both cows continue to lactate. This practice is common in many cattle-keeping societies (Ryan 2005). Although most Maasai would regret losing a calf, the possible loss of the milk from a cow for a full season is a greater loss. Promotion of lactation ranks in the top ten conditions that can be confidently treated by traditional methods (Wanyama 1997).

As calves grow older they join age cohorts that are grazed close to the settlement (Fig. 14.6) or they join with the sheep and goat herds to be herded separately from the adult cattle. Supplementary feeding of calves inside the boma encourages them to eat vegetation at weaning time (Fig. 14.7). Older calves (Figs. 14.8 and 14.9) are grouped to graze further afield, and as each reaches the appropriate age it is introduced to the group where it will learn from the others what is safe to eat and what should be avoided. Calves of any age are never left unattended by a herder, usually a young boy or girl.

Treatment and Prevention of Disease

In treatment and prevention of disease, traditional veterinary treatments are used side by side with western medicines. In an area with so many vector borne diseases, insecticides form a large part of the western and traditional pharmacopeia.

Effective plant derived insecticides include that derived from *Olchilichili* (*Commiphora* sp.). This tree produces a gum like substance on its bark which can be applied directly to wounds or bites, while the roots are boiled in water and the solution is used to wash the animals to prevent tick or flea infestation. *Osukurtuti* (*Cissus quadrangularis* L.) is used in animal bedding, especially in the vestibule of the house where very young calves are housed at night, to prevent fleas or ticks spreading into the house. Western treatments are also used. Cattle are constrained in a temporary corral and sprayed with acaricides. Often the spray does not cover all of the skin. In the Central Kenyan case study described above, many calves were poisoned by an overdosing of the acaracide.

East Coast Fever (Theileriosis) is endemic in many parts of Maasailand today. It is caused by *Theileria parva*, a parasite which spends part of its life cycle in cattle and part in ticks *(Rhipicephalus appendiculatus)*. ECF attacks both young and adult animals and was responsible for many calf deaths in our study areas. After an incubation period of about two weeks, the animal becomes listless, loses appetite, and runs a high fever. Mortality can be as high as 90% in new outbreaks of the disease (West 1985, 255). There is no effective treatment for this disease in either western or traditional veterinary practice although Maasai have several plant derived medicines that they believe help the animal to fight the disease.

To alleviate starvation or dehydration due to drought, traditional treatments are more commonly used. For example, the root of *Olasayiet* (*Withania somnifera*) is boiled, mixed with milk and maize, and given to a calf if its mother dies or is unable to suckle it. It is also given to adult animals during drought.

Conclusions

In the majority of the modern studies we have described, mortality was the result of harsh weather conditions, disease, or neglect but not deliberate culling. Only one Maasai case had a deliberate culling of a calf so that the dam would survive. This is rare, but given the time and effort needed to bring a calf to maturity before it begins to produce calves of its own and give milk, it makes sense to favour an adult cow with a proven record of reproduction. However, given the unpredictable nature of the climate, Maasai are reluctant to slaughter an animal, especially a young animal, that has the potential to thrive if there is a sudden weather amelioration such as the onset of rains.

While all of these studies give important insights into cattle biology and how cattle cope in many different environments, in particular the vulnerability of young calves to disease, they cannot answer specific questions on causes of mortality (culling versus natural causes) in the Scottish Isles during the Neolithic through to the Viking periods. From our perspective, we suggest that the very young calves in the faunal assemblage, those that died in the first few weeks of life, are most likely to have died of starvation, due to the death of their dams or the dam's inability to supply them with sufficient milk. A weak calf may indeed have been deliberately culled to allow another stronger calf to survive by suckling two mothers, or a calf may have been slaughtered to spare the cow the added burden of feeding her calf when she is already weakened by starvation, but the main causes of mortality of cattle of all ages on the Islands may be simply the combined assaults of inclement weather and disease.

References

Binchy, D. A. (ed.) (1978) *Corpus iuris hibernici* i–vi. Dublin, Dublin Institute for Advanced Studies.

Bogucki, P. (1986) The antiquity of dairying in temperate Europe. *Expedition* 28(2), 51–57.

Crabtree, P. J. (1986) Dairying in Irish prehistory: The evidence from a ceremonial center. *Expedition* 28(2), 59–62.

Crabtree, P. (1990) Subsistence and ritual: the faunal remains from Dún Ailinne, Co. Kildare, Ireland. *Emania* 7, 22–25.

Crabtree, P. J. (2003) Ritual Feasting in the Irish Iron Age: re-examining the fauna from Dún Ailinne in light of contemporary archaeological theory. In S. Jones O'Day, W. Van Neer and A. Ervynck (eds.) *Behaviour Behind Bones: The Zooarchaeology of Ritual, Religion, Status and Identity*, 62–65. Oxford, Oxbow Books.

Galaty, J. G. (1994) Rangeland Tenure and Pastoralism in Africa. In E. Fratkin, K. A. Galvin and E. A. Roth (eds.) *African Pastoralist Systems. An Integrated Approach*, 185–204. Boulder, Lynne Rienner Publishers.

Gitau, G. K., McDermott, J. J., Waltner-Toews, D., Lissemore, K. D., Osumo, J. M. and Muriuki, D. (1994) Factors influencing calf morbidity and mortality in smallholder dairy farms in Kiambu District of Kenya. *Preventive Veterinary Medicine* 4, 168–177.

Gusbi, A. M. and Hird, D. W. (1982/1983) Calf mortality rates on five Libyan dairy stations. *Preventive Veterinary Medicine* 1, 105–111.

Halstead, P. (1998). Mortality model and milking, problems of uniformitarianism, optimality and equifinalty reconsidered. *Anthropozoologica* 27, 3–20.

Hamilton, A. C. (1982*). Environmental History of East Africa: A Study of the Quaternary.* London, Academic Press.

Kelly, F. (1997) *Early Irish Farming. A study based mainly on the law-texts of the 7th and 8th centuries AD.* Dublin, School of Celtic Studies, Dublin Institute for Advanced Studies.

Legge, A. J. (1981a) The agricultural economy. In R. J. Mercer (ed.) *Grimes Graves, Norfolk: Excavations 1971–72,* 79–103. London, Thames and Hudson.

Legge, A. J. (1981b) Aspects of cattle husbandry. In R. J. Mercer (ed.) *Farming Practice in British Prehistory,* 169–181. Edinburgh, Edinburgh University Press.

Legge, A. J. (1992) *Excavations at Grimes Graves, Norfolk 1972–1976: Animals, Environment and the Bronze Age Economy.* London, British Museum Press.

Legge, T. (2005) Milk use in prehistory: the osteological evidence. In J. Mulville and A. K. Outram (eds.) *The Zooarchaeology of Fats, Oils, Milk and Dairying,* 8–13. Oxford, Oxbow Books.

Lind, E. M. and Morrison, M. E. S. (1974) *East African Vegetation.* London, Longman.

Marshall, F. (1990) Cattle herds and caprine flocks. In P. Robertshaw (ed.) *Early Pastoralists of South-western Kenya,* 205–260. Nairobi, British Institute in Eastern Africa.

McCormick, F. (1998) Calf slaughter as a response to marginality. In G. Coles and C. Millas (eds.) *Life on the Edge. Human Settlement and Marginality,* 49–51. Oxford, Oxbow Books.

McKay, M. M. (ed.) (1980) *The Rev. Dr. John Walker's Report on the Hebrides of 1764 and 1771.* Edinburgh, John Donald.

McKay, W. M. (1957). Some problems of colonial animal husbandry. The Northern Frontier Province of Kenya. Part I and II. *The British Veterinary Journal* 113(7), 268–79.

Meadows, S. J. and White, J. M. (1979*) Structure of the herd and determinants of offtake rates in Kajiado District in Kenya 1962–1977.* London, Agricultural Administration Unit. Overseas Development Institute.

Mulville, J., Bond, J. and Craig, O. (2005) The white stuff, milking in the Outer Scottish Isles. In J. Mulville and A. K. Outram (eds.) *The Zooarchaeology of Fats, Oils, Milk and Dairying,* 167–182. Oxford, Oxbow Books.

Robertshaw, P. (ed.) (1990). *Early Pastoralists of South western Kenya.* Nairobi, British Institute in Eastern Africa.

Ryan, K. (2005) Facilitating milk let-down in traditional cattle herding systems: East Africa and beyond. In J. Mulville and A. K. Outram (eds.) *The Zooarchaeology of Fats, Oils, Milk and Dairying,* 96–106. Oxford, Oxbow Books.

Ryan, K., Karega-Muñene, K., Kahinju, S. M. and Kunoni, P. N. (2000) Ethnographic perspectives on cattle management in semi-arid environments: a case study from Maasailand. In R. M. Blench and K. C. MacDonald (eds.) *The Origins and Development of African livestock, Genetics, and Linguistics and Ethnography,* 462–477. London, UCL Press.

Sivula, N. J., Ames, T. R., Marsh, W. E. and Werdin, R. E. (1996) Descriptive epidemiology of morbidity and mortality in Minnesota dairy heifer calves. *Preventive Veterinary Medicine* 27, 155–171.

Waltner-Toews, D., Martin, S. W., Meek, A. H. and McMillan, I. (1986a) Dairy calf management, morbidity and mortality in Ontario Holstein herds. I. The Data. *Preventive Veterinary Medicine* 4, 103–124.

Waltner-Toews, D., Martin, S. W. and Meek A. H. (1986b) Dairy calf management, morbidity and mortality in Ontario Holstein herds. II. Age and seasonal patterns. *Preventive Veterinary Medicine* 4, 125–135.

Walter-Toews, D., Martin, S. W. and Meek, A. H. (1986c) Dairy calf management, morbidity and mortality in Ontario Holstein herds. III. Association of management with morbidity. *Preventive Veterinary Medicine* 4, 137–158.

Wanyama, J. B. (1997) *Confidently Used Ethnovetinerary Knowledge among Pastoralists of Samburu, Kenya: Methodology and Results.* Book 1. Nairobi, Intermediate Technology Kenya.

West, G. P. (ed.) (1985) *Black's Veterinary Dictionary.* 15th Edition. London, A & C. Black.

15. A week on the plateau: Pig husbandry, mobility and resource exploitation in central Sardinia

Umberto Albarella, Filippo Manconi and Angela Trentacoste

In this paper an ethnoarchaeological analysis of pig husbandry in central-eastern Sardinia is presented. This research further develops a previous project with a similar focus undertaken in Corsica and northern Sardinia. Results presented here are based on ten interviews with local central-eastern Sardinian herders, and various other observations made in the area. In the study area, the typical Sardinian pig breed is kept mainly free-range, and its style of keeping bears clear relationships with the recent, and possibly more distant, past. Pigs and other livestock are typically kept on a plateau, where they live more or less permanently. Herders regularly move from the lowland villages, where they spend time with their families, to the highlands, where they tend the animals. Though no longer practiced, the long distant movement of pigs – including transhumance – occurred until the 1970s and was aimed at reaching regions where better pig pasturing would seasonally be available. The evidence collected is compared with results from the previous phase of the project, as well as with similar work undertaken by other researchers in Greece and Spain. Mobility patterns of humans and animals have been highlighted as part of this project and their potential for archaeological interpretation is emphasised.

Key words: pig, husbandry, Sardinia, mobility, transhumance

Introduction

The investigation of pig exploitation strategies has produced a number of classic ethnographic studies (Rappaport 1968; Rubel and Rosman 1978; Sillitoe 2003) centred round Papua New Guinea society which have also generated much interpretive interest in archaeology (*e.g.* Redding and Rosenberg 1998). Considering the common occurrence of the species in the archaeological record and the widely accepted archaeological interest in pig exploitation, it is surprising that ethnoarchaeological analysis has, until recently, paid little attention to the subject outside the Pacific area. However, numerous projects in the last decade have sought to specifically address the ethnoarchaeology of pig exploitation. Observations that we carried out in the 1980s and 1990s on the islands of Corsica (France) and Sardinia (Italy) and the following more structured fieldwork project undertaken there in 2002, have contributed to filling this knowledge gap (Albarella *et al.* 2007). In more recent years, work done by Halstead and Isaakidou (in this volume) in Greece has contributed a similar perspective on the eastern Mediterranean area, using a different approach but sharing similar research interests. Hadjikoumis' doctoral dissertation (2010) on pig domestication in Spain extended the methodological approach that we originally adopted in Sardinia and Corsica to the Iberian Peninsula, focusing on the famous *dehesa* environment of Extremadura (Spain) and neighbouring Portuguese areas. A similar approach was used by Scrivener (2010) in her undergraduate dissertation focused on a pilot study of pig exploitation in the New Forest of southern Britain. In this area some practices are reminiscent of the woodland pig pasturing – *pannage* – which was widespread in England during the Middle Ages (Wiseman 2000; Albarella 2006). No longer a neglected topic, the ethnoarchaeology of pig husbandry has, in other words, recently become a very active and vibrant area of investigation.

This paper aims to be a further contribution to this area of study, and should be considered an extension of our original work in Corsica and Sardinia (*cf.* Albarella *et al.* 2007). In that first stage of the project we only marginally touched upon central Sardinia but were aware that husbandry practices in this area have remained the closest to their historical roots. At the same time, the wildness and imperviousness of the countryside have helped to preserve the original biological characteristics of wild boar and domestic pigs, as well as traditional aspects of the life of local communities. The key aim of this second phase of our project was therefore to extend our research into this geographic area in order to:

- Increase our sample of interviewed herders, as an aid to check the reliability of our previously collected evidence, as well as to enhance its geographic coverage;
- Investigate possible differences in husbandry practices between different areas of Sardinia;
- Explore our working hypothesis that central Sardinia was more culturally isolated than the northern part of the island, and therefore less affected by the introduction of husbandry innovations which originated on the nearby continent.

This paper presents the results of this investigation into central Sardinian pig husbandry, and discusses the implications of this research in the context of previous ethnoarchaeological work done in Sardinia and other geographic areas, with special attention paid to the aspects that have the greatest potential for archaeological interpretation.

Study Area and Methods

The evidence discussed in this paper derives from interviews with ten swine herders (Table 15.1); a number of additional conversations with various other local people; fieldwork observations; and photographic recording of the landscape, environment, architectural features related to pastoralism, and the animals themselves. Although the results are informed by many years of data collection in the region, they mainly refer to fieldwork undertaken between August and September 2005. The sample of interviewed herders is not statistically representative, but, particularly if used in combination with the other eight interviews we carried out in southern Corsica and northern Sardinia in 2002 (Albarella *et al.* 2007), should nonetheless be informative. It is also important to consider that we actively biased our sample towards breeders who kept their animals free for at least part of the year, as we considered these to have a greater potential for addressing archaeological questions. Pigs that are enclosed around the year do, however, also occur, though they are less common and nowadays are generally represented by imported rather than local breeds.

The area concerned by this paper is located in the central-eastern part of Sardinia (Italy) (Fig. 15.1) in the administrative province of Nuoro, and more specifically in the geographic areas known as Supramonte (herders A and B), Ogliastra (herders C to G) and Barbagia (herders H to J). Geologically the area is, like most of Sardinia, rather diverse, and includes limestone formations (towards the coast), as well as granite and metamorphic rocks. It is a mountainous landscape with variable elevations ranging from sea level to above 1800m in the Gennargentu Mountains, on whose slopes some typical free-range livestock are kept. The vegetation is typically Mediterranean with substantial pockets of surviving woodland (holm oak, cork oak, downy oak, sweet chestnut) as well extensive areas of maquis.

The pig herders involved in interviews were chosen from several villages with various local landscapes. The two Supramonte herders (herders A and B) were interviewed in and around the village of Dorgali (400m asl), which is only about 10km away from the coast. Both herders keep their pigs in the hills between the village and the coast: one in an area of dense woodland (A) and the other (B) in a slightly more open landscape, centred around a remarkably well preserved shepherd's hut (Fig. 15.2), inside which our interview occurred. Although swine flu had recently devastated the pig herd of this breeder (B), at the time of our visit the area around the hut was teeming with roaming pigs and some enclosed piglets were also noticed. In Ogliastra, we interviewed three herders from the village of Urzulei (511m asl) and one from Villanova Strisaili (845 m asl), though the former kept most of their pigs on higher ground above the village. Urzulei was, according to many of the people to whom we spoke, considered to be the most typical 'pig territory' in the whole of Sardinia, and roaming pigs can often be seen while driving in that area. The location of the final herder (G) we interviewed in Ogliastra is noteworthy, as his pigs were located on the limestone plateau of San Pietro (385m asl), near the village of Baunei. This is a remarkably wild and isolated

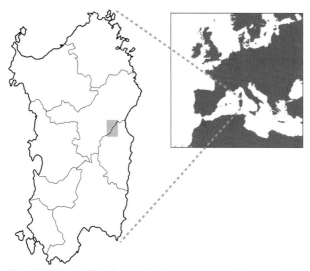

Fig. 15.1 *Map of Sardinia with the approximate location of the study area highlighted. Compiled by A. Trentacoste.*

Fig. 15.2 *A typical shepherd's hut (pinneta) at Buchi Arta (Dorgali). Photo by U. Albarella.*

area, which we first visited in 1986; it was reassuring to see that after almost 20 years the area still hosted thriving populations of free-range pigs. A small church and a restaurant are the only human-made constructions visible on the plateau. Finally, we also conducted interviews with herders from Orgosolo (590m asl), in Barbagia. This is the most inland location that we chose, and it provided us with the clearest indication of the 'plateau style' of livestock management that will be described in the next section.

Information from the herders was obtained through a questionnaire composed of about 25 questions and slightly modified from the one which was used in our 2002 fieldwork. Although much information was also gathered outside the more formalised question/answer sessions, we considered it important to have a more structured component to the conversation so that a similar type of information could be compared between different herders. Handwritten notes were taken of all answers provided by the herders and, in addition, all conversations were tape-recorded.

Results

A Week on the Plateau

Before we delve into the details of the information provided by the herders (summarised in Table 15.1) it is useful to draw a general picture of the style of livestock husbandry that characterises our study area, as illustrated by conversations with the interviewees and other informers.

Until recently, in central-eastern Sardinia, it was typical for people to use a more permanent settlement (village) in the valley and a temporary shelter on the plateau, where livestock would be tended. Still today, the men of the village who own livestock (herders) periodically move from one location to the other, while the livestock – comprising of cattle, sheep and goats in addition to pigs – stays on the plateau. This system is exemplified by the Orgosolo case study, where the herders used to walk about 12km to a plateau located almost 500m higher in the mountains (Fig. 15.3), in order to spend time dealing with the animals.

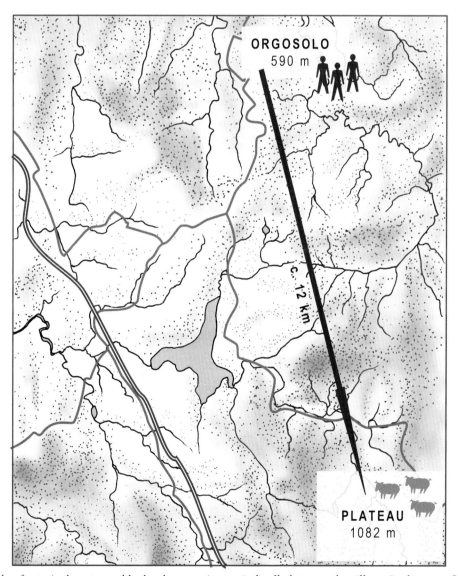

Fig. 15.3 Example of a typical route used by herders, moving periodically between the village (in this case Orgosolo) and its plateau. The animals remain on the plateau. Compiled by A. Trentacoste.

Table 15.1 A summary of the results of the interviews of pig breeders carried out by U. A. and F. M. in August and September 2005. All breeders are based in central/eastern Sardinia in mountainous areas. They are all located in the administrative province of Nuoro and the geographic areas known as Supramonte (A, B), Ogliastra (C to G) and Barbagia (H to J). Continued over the next three pages.

Locality	Buchi Arta (Dorgali)	Dorgali	Urzulei	Urzulei	Urzulei	Villanuova Strisaili	S.Pietro plateau (Baunei)	Orgosolo	Orgosolo	Orgosolo
Breeder	A	B	C	D	E	F	G	H	I	J
Herd size and composition	currently none due to devastation by swine fever; previously 40–60. In the past specialised swineherds could have as many as 200 pigs	c. 20, but up to 100 in the past. Herd is a mix of sexes and ages	c. 30, but up to 350 in the past	c.15 but up to 150 in the past. Herd includes a boar, a sow with piglets, and castrates	c. 50. In the 1960s herd was <30 due to scarcity of feed. The herd is divided into two groups: one kept in the village and the other on the plateau. Herder keeps 4 entire boars	c. 40, which is the average today. In the past some breeders had hundreds of pigs. Slaughtered suckling piglets are mainly females, while the adults are mainly castrates	c. 30; herding is in addition to managing a restaurant. In the past specialised pig breeders were common	c. 40; in the past specialised pig breeders could have as many as 400 animals	in the past 300–400; castrated males were the majority	28 sows, 3 boars and 12 piglets
Breed	traditional Sardinian, with occasional accidental crosses	traditional Sardinian (NB: the purest possible!) kept free range. Herder also owns improved pigs, but they are enclosed as they would not survive on the plateau	pure traditional Sardinian	traditional Sardinian, but with some crosses	pure traditional Sardinian	mainly traditional Sardinian, with a few improved animals	pure traditional Sardinian	traditional Sardinian	traditional Sardinian	traditional Sardinian
Any wild boar also kept?	occasionally hybrids	some in the past	no	no	2 wild boar + 11 in the mountains; wild boar are kept for reintroduction in the wild (but males are slaughtered)	no	no	no	no	one wild boar
Any wild/domestic crosses?	it happens; the hybrid is wilder but manageable. Hybrid is sometimes killed; other times allowed to grow	it happens regularly, but the hybrids are immediately slaughtered because they do not grow enough (the max weight for the wild boar is 60kg)	it happens, but they are killed	not recently, but wild boar do live in the area	it happens	yes. Hybrids are generally slaughtered because they don't grow enough and tend to escape	it happens. Hybrids are slaughtered because they grow too slowly	yes, it is common, but they are generally slaughtered because they do not grow enough	crosses have happened, and even though small, they were still kept	it happens and the herder owns a hybrid. He kills the males and keeps the females although aware that they will not grow as much as the other pigs
Other livestock/activities	goats, cattle, vineyard	sheep, goats, cattle, olive grove	goats, dairy products, hay meadow	none currently, but it is family tradition to keep also goats (for milk and meat)	cows (for family-based milk use), donkeys, horses, vegetable garden	cattle are kept for meat, and sheep for dairy products. Sheep meat and wool are of little value. Goat milk is also used in the area	goats (meat, milk), sheep (meat, milk), cattle (meat), chickens	sheep, goats, cattle; specialised pig breeding is a thing of the past	sheep, goats, vegetables, vineyard	goats, cattle, donkeys
Castration	90–95% are castrated at 3–4 months; castration is very safe	almost all males are castrated when about 1 year old; no problems are associated with castration	castrated when 1 or 1.5 years old; only one male is kept entire for reproduction	almost all males are castrated at 7–8 months	90% males are castrated at 5 months. Boars kept for reproduction may be castrated at 2–3 years when they will be replaced by a younger male; beyond that age boars may become dangerous	castration occurs between 3–6 months; all males are castrated except the best specimens; no problems are associated with castration	castration is generally at 5–6 months, although boars may be castrated at 2–3 years	castration normally occurs at c. 6 months. Most animals are sold as piglets, but some (c. 10) are castrated and kept for family use	almost all males are castrated	castration is done in August and the age is very variable

	1	2	3	4	5	6	7	8	9	10
Birth season and litter size	generally Dec–Feb, but sows can give birth twice a year. Up to the 1950's one birth per year was the norm due to poorer nutrition. With intensive feeding sows can be pushed to give birth three times per year	mainly in autumn, but other times are possible; sows can farrow twice a year, but normally do so only once; the litter is composed of 5–6 piglets, but in the past this could be as low as 3	can happen throughout the year; generally sows farrow twice per year	piglets can be born at any time of the year; sows have two litters per year; litter size is 8–10 piglets	can happen throughout the year; the purest Sardinian breed generally farrows once a year, but it can do so twice if kept inside; the litter is composed of c. 7 piglets	throughout the year, twice a year; litter of 9–10 piglets, but in the past half of those were killed	throughout the year, twice a year, but occasionally once and even not at all according to the season and the health of the animal	throughout the year, every four months (three farrows per year); c. 12 piglets per litter	twice a year; litter generally of c. 8 piglets but nowadays can go up to 12	twice a year, litter of 8–10 piglets
Mating season	females are in oestrus when piglets are removed; generally twice a year, most commonly in Sept–Oct	any time	throughout the year; 4–5 days after the piglets are removed the sow goes again in oestrus	throughout the year	throughout the year	throughout the year	throughout the year	throughout the year	throughout the year	throughout the year
Where are the litters born?	either in the pen or the countryside; if the latter, there are more losses due to predation from foxes and hawks	they are born outdoor but generally under shelter; since piglets are often predated by foxes or hawks they can be kept in the sty for as long as three months	they are normally born in the sty, where they stay for 15–20 days	either in the pen or the countryside; if the latter, there are more losses due to predation from foxes	generally in the sty, but occasionally outdoors where they are prone to fall prey to foxes; the piglets leave the sty after c. 20 days	in the pen; piglets need to be protected from other pigs; after 20 days the piglets can leave the pen	they are generally born in a pen	they are generally born in a pen; after c. 40 days they are let out	generally in the sty; after c. 15 days the sow takes the piglets outdoors	mainly in the sty, but they leave after 10 days
Purchase of animals	no; unless, like now, the herd dies out due to disease	no	no	no	no; but the herder lends his boar to others; he may also borrow one from others if this is not in the area	no; the herder once purchased pigs from the Campidano valley, but they did not survive		only occasionally	no	only boars for reproduction
Age at slaughter	castrated males are normally slaughtered between 1–2 years; slaughter of suckling piglets is not a traditional practice, but nowadays they are killed at 40–50 days (6–7kg)	c. 1.5 years; suckling piglets at 1–2 months (6–7Kg)	if the aim is ham production slaughter occurs between 1.5–3 years; suckling piglets are killed when 4–5kg	castrates at c. 1year; boars at 2–3years; suckling piglets at 30–40days (c. 9kg)	generally when 1 year old; suckling piglets at 35 days (8–9kg)	generally when 1 year old; suckling piglets at c. 40–50 days (7–20kg)	the breeder mainly slaughters pigs as suckling pigs for the restaurant (1–2 months, c. 10kg)	suckling piglets normally at 40 days, castrated animals (for family use) are normally slaughtered at 1.5 years	suckling piglets at 40–50 days; otherwise when c. 1 year old	suckling piglets at 40–50 days
Slaughter season	Jan–Feb when the pigs are fattened and their meat preserves better	winter, when the meat can be more easily preserved	Jan–Feb	Dec–Feb, when low temperatures are more suitable for ham preparation	Dec–Mar, when pigs will have fattened themselves with acorns and the meat is also tastier; cooler temperature also guarantees better preservation of the meat	Dec–Feb when the meat can be cured without the risk of spoiling. Pig meat is believed to last longer than other types of meat outside of a refrigerator	winter, when the temperature is right to cure the meat	Dec–Jan, when it is cold and the meat preserves better	winter, when it is colder (Nov–end of Mar)	during the Christmas holidays
Enclosure and home range of the pig herd	entirely free but the pigs decide to stay within a relatively limited area	entirely free; kept at least at 800m asl	entirely free	the pigs are generally enclosed, but in the past they were kept outdoors all year	entirely free on communal land; pigs can go as far as 10km away but they always come back	entirely free but they tend not venture too far – except entire males that disappear sometimes for months	entirely free in winter but they are enclosed in summer, generally Mar–Nov	entirely free on communal land	entirely free on communal land	they live in a area of c. 60 hectares

Locality	Buchi Arta (Dorgali)	Dorgali	Urzulei	Urzulei	Urzulei	Villanuova Strisaili	S.Pietro plateau (Baunei)	Orgosolo	Orgosolo	Orgosolo
Breeder	A	B	C	D	E	F	G	H	I	J
Daily movements	pigs tend to return to the pen for the night, but in summer they often spend the night in the countryside; sometimes they find shelter in caves	pigs return to the pen at night, both in winter and summer. If they do not, they are either unwell or the pen is inadequate	pigs are normally outside, especially in areas where water is available	NA	pigs return to the sty at night when it is cold, otherwise they stay outside. They can also find shelter in caves and hollow trees	both in winter and summer pigs tend to stay outside, though they may go back to the sty if this is very comfortable; they find various areas of natural shelter	pigs return to the pen for the night	only in winter do pigs return to the pen at night	pigs return to the pen at night; during the day they are generally found near water sources	pigs return to the sty for the night, but only in winter; they can find shelter in the rocks and scrubs. In summer they generally pasture along streams
Mobility	in the past transhumance was practiced, mainly towards the Campidano plains where the pigs would often been sold; in bad years for acorns the animals might be moved to areas where acorn production had been better, such as the Limbara or the Desulo and Aritzo areas			in the past (up to the early 1960s) the pigs were moved from the mountains to the Campidano valley or other plains where they could feed on stubble. They would normally be taken back in Oct/Nov. The breeders themselves took them to the plains. Up to 100-150 pigs could me moved by two or three breeders (in later years on the train!)	the pigs mainly stay on the plateau and the breeders could spend up to 20-30 days with them away from the village; in the past in summer people used to take their pigs to the Campidano valley to feed on stubble, but some would go to closer locations	in the past pigs were taken as far as Cagliari to be sold, but in Jul and Aug they would also be taken to the Campidano valley to feed on stubble	historical transhumance towards the Campidano is known for this area as well	up to the 1960s and 1970s, in summer the pigs were taken to the plains around Cagliari and Oristano to feed on wheat and barley stubble; they would remain in that area for about 3 months. Sheep, goat and cattle transhumance is still practiced today	in summer the pigs could be taken more than 100km away to the Campidano valley, where they would feed on stubble; in bad years for acorns, pigs would be taken to the areas of Aritzo and Desulo where they would feed on sweet chestnuts	
Level of control	minimal; pigs are completely free, mainly independent; but in summer pigs require to be fed daily	minimal, pigs are completely free, but need to be fed in summer; in winter the breeder may not see the pigs for as long as a month	minimal; pigs are completely free, but need to be fed in summer	NA	minimal; the pigs are free, and only occasionally enclosed; the breeder may not see the pigs for as long as 2–3 months	minimal	minimal in winter	sows with piglets require quite a lot of attention, but those that are free are only checked a couple of times a week (slightly more frequently in summer)	minimal in winter, but in summer pigs must be fed daily	minimal as they are kept free range
Capture for slaughter	attracted by food; occasionally shot	attracted by food	attracted by food	NA	pigs are generally attracted by food, but in winter when there are many acorns around it may be difficult to attract them; occasionally they need to be shot	pigs are attracted with food to a pen	NA	call from the breeder	call from the breeder	pigs voluntarily approach the breeder
Diet	in winter: mainly natural food (acorns) with occasional supplements (maize, buttermilk) provided by the breeder. In summer: mais and barley provided by the breeder	in winter (Nov-Feb) pigs are entirely self-sufficient (acorns, tubers, worms), but in summer they are supplied with corn and barley everyday, though they also find some of their own food; even in summer if there are no piglets around daily feeding may not be needed	acorns in winter; only sows that have just given birth need some food supplementation; corn seeds are supplied by the breeder in summer.	breastfeeding sows are given concentrate, while all other pigs are fed with corn seeds throughout the year. In winter this practice is supplemented with some acorns that the breeder gathers in the woodland	in winter they need no supplements to natural food but some is provided to prevent pigs becoming too wild. In summer there is little food around and they need to be fed with corn seeds. In the 1960s in the summer the pigs survived on figs, prickly pears and stored acorns	in winter pigs are completely self-sufficient (acorns and some worms); in summer they feed on grass, roots and worms, which is supplemented by corn and legumes provided by the breeder	unless there has been a bad year for acorns, pigs are self-sufficient in winter; in summer they need to be fed with corn and legumes	in winter pigs feed on acorns and grass and they are entirely self-sufficient (expect sows that have just given birth); in summer pigs are supplied every day with maize and barley which supplements the rather scarce natural food	acorns, roots, worms	in autumn/winter pigs are self-sufficient and eat acorns, grass, tubers, but a food supplement is provided to make sure they remain accustomed to people. In summer they eat roots, but this is supplemented with maize and wild pears provided by the breeder

Adult Weight	full growth: max 130kg; at 1 year average 55kg, max 90kg; wild boar max 80–90kg	full growth at about 2.5 years (max 100kg but average is 70kg)	full growth: average 90kg, max 120kg	traditional breed: max 100–120kg; max growth reached at 1 year	at 1 year max 100kg; at 2 years max 130kg	100–150kg when 2 years	max 150kg when 2 years, but the purest Sardinian breed is lighter	suckling piglets: 7–8kg max weight at 2 years is 200kg	max 100kg	suckling piglets: 6–7kg; adults range from 80–200kg, but the average is c. 150kg. Pure Sardinian wild boar only eat grass and insects and weigh no more than 25kg (!)
Losses	rare (occasionally stolen)	occasionally pigs disappear, especially entire boars		NA	occasionally (4–5 per year)		rarely, generally entire boars	rarely	the adults are sometimes stolen	there may be casualties due to hunters or fights between males
Agricultural damage	none; no snout wire needed	there is no cultivated land around, so no snout wire is needed		NA	iron wire is used in the snout to avoid crop damage and deter the pigs from moving too far; however they become fatter without the wire		no; cultivated fields are fenced. Others use iron wire in the pigs' snouts but this breeder does not, as otherwise the pigs cannot feed properly (they can only graze)	not at all due to absence of agriculture on the plateau; snout wire is not necessary	not an issue due to the absence of cultivated fields	rarely a problem and no snout wire is used, except for entire boars to prevent them from killing sows and piglets
Products	meat, fat; in the past bristles were also used and the meat was sold with the skin still on; skin was also used	meat; but in the past (up to the 1960s) bristles were also used and the skin was used to make shoes and bags	in the past living animals were also sold for families to fatten up	meat (also cured), fat	meat, and occasionally skin to produce bags and shoes; however pigskin is not considered to be very good for this purpose		meat, fat; in the past bristles were mainly used by cobblers and to make brushes	meat, blood	meat, fat; in the past bristles were used by cobblers	meat, fat; in the past bristles were used for making brushes, and the fat was the most valuable part of the carcass	only meat

In the pre-motorization days it was unfeasible to cover this distance and altitudinal range every day, and as a consequence, the herders would spend extended periods of time (a week or more) on the plateau, finding shelter in purposefully built huts, known locally as *pinnetas*. These constructions still represent a common feature of the central Sardinian highlands (Fig. 15.2), where simpler and smaller structures known as *sarule* (Figs. 15.4 and 15.5), devoted for the shelter of animals rather than people, can also be commonly encountered. The movement of men between the village and the plateau was not at all seasonal, but would rather proceed unchanged throughout the year.

The solitary life of the men on the plateau has often contributed to the formation of tribal associations among family groups. This practice has given rise to the mythology associated with the notorious banditry of the Sardinian shepherds, but also to the creation of a strong cultural identity and a pride for the historical as well as natural heritage of the region. These themes all feature prominently in the murals that decorate the walls of many houses in Orgosolo (Fig. 15.6), and have amply been portrayed in both literature and cinema.

The central Sardinian plateau is generally characterised by areas of woodland combined with rather extensive open areas (Figs. 15.7 and 15.8). In the latter, shrubs are prevented from growing excessively by the continuous grazing and browsing of livestock. By keeping the vegetation low the animals fulfil an important ecological role, as they help in preventing the spread of wood fires, an endemic problem in the region. The plateau is almost invariably made of communal land, which makes this style of husbandry possible, as the animals are free to pasture wherever they like. The virtual absence of agriculture on these high grounds means that there is no clash of interest between farming and herding, and that crop damage caused by livestock is generally not an issue. Large scale agriculture is not reported in this area, and around the valley bottom village people more commonly engage in horticulture, accompanied by the cultivation of small plots of land and the keeping of a few animals, generally – unlike on the plateau – kept enclosed. Local communities are remarkably self-sufficient and even nowadays in some villages there are only a few shops. Much of the foodstuff required by a family is acquired through exchange rather than purchase and homemade production is still the norm.

The advent of the car has substantially changed the mobility pattern, making it feasible for herders to move from the village to the plateau on a daily basis. However, this is a surprisingly recent development, as car ownership has, in this area, only become commonplace in the last four decades. Although we did not have the opportunity to interview them, we were told of a few elderly shepherds who still regularly spend the night on the plateau, particularly in the Supramonte area. Although many *pinnetas* have fallen into disuse, others are still remarkably well preserved, sometimes as heritage monuments, but more often because they still play a useful function as storage rooms and resting places. By and large the general style of husbandry has

Fig. 15.4 A typical animal's shelter (sarula) in the Orgosolo area. Photo by U. Albarella.

Fig. 15.5 A sarula with pigs in the Orgosolo area. Photo by U. Albarella.

Fig. 15.6 Mural painting from Orgosolo representing a shepherd wearing traditional clothes. Photo by U. Albarella.

remained unchanged to today, except that the movement between the village and the plateau has become suitably faster.

Herd Size and Structure

Large pig herds (up to 400 animals) of the kind still found in the Spanish *dehesa* (Hadjikoumis 2010) are mainly a thing of the past in Sardinia. In the past pig keeping was sufficiently profitable for people to specialise in pig husbandry and consequently herders owned many animals, but this is no longer the case. Nowadays most pig herders also have other professions and/or keep other livestock. For the past four decades it has been more manageable to keep small herds of 20 to 40 pigs. The herd normally includes one to three boars (occasionally none, which means that a boar has to be hired), a few piglets and a majority of castrated males. One of the herders (F) reported that most females would be slaughtered as suckling piglets, while most of the adults would be castrated males. However, this herd composition is not universal as breeder J had a majority of adult females. Most breeders are self-sufficient and only very rarely buy pigs from elsewhere.

Fig. 15.7 A view of the Orgosolo plateau. Photo by F. Manconi.

Breed

All the interviewed herders owned pigs of the traditional Sardinian type, which is dark-coloured, slim and long-snouted (Figs. 15.9 and 15.10). These pigs are more similar to wild boar than to improved pig breeds, mirroring the morphological types characteristic of free-range husbandry

Fig. 15.8 One of the authors taking pictures of pigs on the Orgosolo plateau. Photo by F. Manconi.

Fig. 15.9 A typical long-snouted pig of the Sardinian 'breed' from the Urzulei area. Photo by U. Albarella.

in Corsica (Albarella *et al.* 2007), Greece (Halstead and Isaakidou, in this volume) and Spain (Hadjikoumis 2010). Although one of the herders (D) also had some local pigs crossed with other breeds, the situation in central Sardinia in clearly different from the one we witnessed in the north of the island, where improved breeds had introgressed the local types much more heavily (Albarella *et al.* 2007). Occasionally, improved pig breeds are also kept, but these are enclosed since they would not survive free-range life on the plateau, therefore confirming the accounts that we also received in Corsica. Even pigs acquired from the Campidano valley in south-west Sardinia (Fig. 15.11) can fail to survive on the plateau (Table 15.1). Some of the herders rather proudly claimed to own a very pure Sardinian breed, though genetic analysis that we carried out on the hair of the pigs of breeder B (Larson *et al.* 2007) revealed that, as far as mitochondrial DNA is concerned, they were indistinguishable from the main genetic type widespread in Europe. This is not surprising, as the Sardinian pig type, like the Corsican (Porter 1993, 135), represents a combination of populations of rather diversified origins, rather than a genuine breed. Its rusticity and morphological distinctiveness, reminiscent of the pigs described by Cetti and della Marmora in the 18th and 19th centuries respectively (Albarella *et al.* 2007), cannot, however, be questioned.

Wild Boar

The wild boar is not an endemic species in Sardinia, but it has been found there since the Neolithic. It was either introduced by early farmers or, more likely, it originates from early domesticates that became feral (Albarella *et al.* 2006). The species is widespread in the region but there have been many introduction events of wild boar from the continent that have diluted the purity of the Sardinian populations. Our study area, however, and Ogliastra in particular, is regarded to be one of the areas where introgression has been more limited (Onida *et al.* 1995),

Fig. 15.10 A typical dark and slender pig of the Sardinian 'breed' from the Dorgali area. Photo by U. Albarella.

perhaps because it is a traditional pastoral area with limited agriculture (*cf.* Cetti 1774, 110). The Sardinian wild boar represents one of the smallest sized populations of the species across its range (Albarella *et al.* 2009) and this has important consequences for its management. Some informers mentioned a maximum weight of 60kg for the wild boar, but also reported the occurrence of 25kg (!) animals, which would only eat grass and insects. However unlikely such weight may seem, it is not far from the lowest end of the weight range provided by Toschi (1965, 429) for the Italian wild boar (*i.e.* 30kg). Only occasionally pig herders keep wild boar, and when they do it is generally in very small numbers. One of our interviewees (breeder E) bred wild boar with the aim to reintroduce them into the wild for hunting purposes – the males would, however, be slaughtered. Our informants rather consistently agreed that domestic pigs and wild boar would commonly mate with each other. This is a well-known phenomenon, already documented in the 18th century (Manca Dell'Arca 1780,

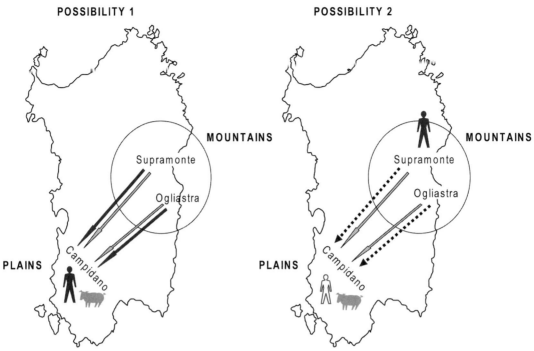

Fig. 15.11 A representation of the seasonal movement of pigs and people from the highlands of Ogliastra and Supramonte to the Campidano valley. In the first possibility the pig owners travel with the pigs, while in the second possibility the breeders hire seasonal workers to drive their pigs to the lowlands. Compiled by A. Trentacoste.

323). Hybrids are occasionally kept (Fig. 15.12), but in general they are not welcome as they do not grow large enough to produce sufficient meat and, in particular, they grow slowly. Consequently hybrids are generally slaughtered early, particularly the males. This approach is consistent with the evidence gathered in Corsica and northern Sardinia (Albarella *et al.* 2007), Greece (Halstead and Isaakidou in this volume) and Spain (Hadjkoumis 2010).

Castration

Castration is a routine practice, as it was in the 18th century (Manca dell'Arca 1780). Like in Corsica, northern Sardinia (Albarella *et al.* 2007) and Greece (Halstead and Isaakidou in this volume), but unlike Spain (Hadjikoumis 2010), castration is only practiced on males. About 90–95% of the males are generally castrated but, as in Greece, the age of castration is rather variable. Most informers indicated that castration occurs when the animals are between three and eight months old, but castration occurring in the first half of the second year was also mentioned. The reproductive boar is generally kept until it is two to three years old and then castrated and replaced by a younger animal. Boars older than several years are regarded to be potentially dangerous, and in fact fights between males may even lead to casualties. This timing is by and large consistent with the advice provided by Roman sources, for example in the first century AD Columella (VII.9.5) wrote that castration should occur when the pigs are six months old and the reproductive males are three years old. At any age the practice of castration is regarded to be safe and no casualties were reported. Only one breeder mentioned this to be a seasonally activity, generally occurring in August.

Fig. 15.12 A young wild boar x domestic pig hybrid from the Urzulei area. Photo by U. Albarella.

Life Cycle

Female pigs are apparently in oestrus throughout the year, except when they are lactating. They seem to go back to oestrus only a few days after the piglets are removed. The consensus is that sows generally farrow twice a year, but this requires them to be fed adequately. In case of food shortage the traditional Sardinian pig will give birth only once a year, and in extreme cases, not at all. This is very similar to the situation that has been described for Greece (Halstead and Isaakidou in this volume). Additionally, treble farrowing was mentioned as a possibility by a couple of our informers, provided that the pigs were particularly well fed. However, according to breeder A (perhaps significantly, the eldest man we interviewed) one litter per year was the norm up to the 1950s. Apparently the piglets can be born at any time of the year, though two of the herders mentioned autumn and winter as the preferred birth seasons.

The overall health and level of nutrition of the sow play a role in the size of the litter. Litter size is variable, and numbers as low as five and as high as twelve were mentioned; the most common figures seems to range around eight to ten piglets. In the 18th century it was accepted that a sow could milk no more than five or six piglets. Therefore if more were born, they were killed and eaten as a delicacy, and in cases of shortage of food, all would be slaughtered to prevent the mother from eating them herself (Manca Dell'Arca 1780, 324). Two breeders provided a more recent historical perspective: one by making reference to the fact that in the recent past, a litter of as few as three piglets would not be unusual; the other by mentioning that litter size had increased over time from an average of about eight piglets to up to twelve. This variability is also consistent with the wild form, which gives birth to between three and twelve piglets (Toschi 1965, 433).

The sow normally gives birth in a sty or pen, but births in the countryside are not unusual, and herder F showed us a hollow oak tree where, until the recent past, sows would commonly find shelter when in labour (Fig. 15.13). Births outside a protected area are avoided if possible, as many piglets are predated by hawks and foxes. These carnivores are particularly feared and disliked by the breeders who kill them when the occasion arises. On the plateau above Orgosolo we were taken to a rather creepy spot where several dead foxes had been hanged to trees (Fig.15.14) in the belief that this would scare their living counterparts away. When born in a pen or sty (or better a *sarula*), the piglets are kept enclosed for 10 to 40 days (15 to 20 days representing a more average figure). Herder B, however, mentioned keeping the piglets inside for as long as three months, which would be consistent with the weaning age known for the wild boar (Toschi 1965, 433). Herder B was one of our most accurate and reliable informers, who keeps very pure Sardinian pigs, and his estimate, despite being at odds with others, is therefore credible.

Diet

As mentioned, in autumn and winter pigs are by and large independent as they can freely feed on acorns, tubers, grass and worms. Bad years for acorn production present an exception to this pattern; the way these were dealt with will be discussed below. In winter additional food is occasionally provided by the herders, but several herders

Fig. 15.13 A hollow oak tree where sows used to find shelter in the area of Villanova Strisaili. Photo by U. Albarella.

mentioned that the main purpose of this intervention is not nutritional, but is rather aimed to prevent the pigs from becoming too wild and unaccustomed to human presence. Winter food integrations that were mentioned included maize and whey (Fig. 15.15). Suckling sows are helped at any time of the year and are often fed on food concentrate. In summer there is much less natural food available and the pigs are consequently more substantially supported. They still find some food around, such as grass and roots, but this is heavily supplemented with corn seeds (mainly maize and barley), legumes and occasional fruits (pears) provided by the herder. Some of this food is produced industrially and was not available in the past. One of the herders (E) mentioned that in the 1960s pigs survived the summer on supplied figs, prickly pears and stored acorns. This was also confirmed by an elderly farmer, with whom we spoke, who kept his pig enclosed all year round and fed it with barley from his fields. Significantly, he said that the pig could not be fed on food scraps, because the family would not leave any; in those much more parsimonious days everything would be eaten thoroughly.

Weight

Most herders mentioned that Sardinian pigs reach their full

Fig. 15.14 Hanged fox corpses on the Orgosolo plateau. The animals were killed by herders in the belief that they would scare away their living counterparts. Photo by U. Albarella.

Fig. 15.15 A pig feeding on whey in the area of Villanova Strisaili. Photo by U. Albarella.

growth when they are about two years old, though one of them (D) said that this occurred at one year and another (B) at one and a half years. The variability is likely to be due to feeding regimes as well as to the purity of the breed, with the most typically unimproved Sardinian pig probably being rather slow growing. At full growth the maximum weight of the living pig is given as ranging between 100 and 200kg, but the average can also be below 100kg. To give an idea of the growth rate two breeders mentioned that at one year of age, the average pig weight would be 90kg (A) or 100kg (E), whereas the following year this would have increased to 130kg. There was consensus around the fact that the purest Sardinian type would, on average, be lighter. The elderly farmer said that, despite keeping it enclosed, his typical Sardinian pig would not grow beyond 100kg. These figures compare well with those provided for unimproved breeds in Corsica and northern Sardinia. In central Sardinia the heavier (up to 300kg) improved pigs, which are occasionally found in northern Sardinia, do not seem to occur.

Products

Meat and fat are the key products of the pig. In the past – when dietary priorities were clearly different from modern days – lard in particular was regarded to be the most valuable part of the carcass. The meat is mostly cured to produce hams, some of which have a very high proportion of fat. However, the taste is, as we experienced in Urzulei, delicious and entirely different from that produced through intensive breeding. Among other products mentioned, blood was also regarded to be useful. Pigskin was in the past used for the production of shoes and bags, but this practice has now been abandoned. Pigskin was never regarded to be particularly valuable, but carcasses were often sold unskinned, as the skin could also be used for culinary purposes. In the past (until the 1960s) pig bristles were used for making brushes and by cobblers, a fact also reported by Manca dell'Arca (1780, 326). Finally, it was also mentioned by herder C that young pigs could also be sold alive to families who would fatten them up within the household. This practice, however, also seems to have disappeared. Presumably this is because nowadays families tend to acquire improved animals that would grow larger and more quickly.

Slaughter

Adult animals are slaughtered between one and three years of age, with castrates being killed towards the younger part of the range (second year) and breeding pigs towards the older (third year). These pigs would mainly be used for ham production. In Sardinia, however, a fashion for the consumption of meat of suckling/very young animals has developed in recent decades, but this is not a traditional practice, as we were also informed in the north of the island (Albarella *et al.* 2007). These young pigs are normally killed when one or two months old (the figure of 40 to 50 days was repeated by several breeders) and with a weight ranging between four and ten kilograms. As in Spain (Hadjkoumis 2010) and in the north of the island, the slaughter season of the older pigs is firmly fixed in winter, when the animals are fatter after pasturing in the forest and the meat is less prone to go off in the colder climate. Additional advantages of killing the pigs in this period include the potential coincidence with the Christmas holidays (herder J) and the fact that the meat is tastier after the pigs have spent time feeding on acorns, which become available in autumn (herder E).

Territory and Control

All the interviewed herders generally keep their pigs entirely free-range, though this has become a challenge in recent years due to regional laws which, to prevent the spread of swine flu, restrict the movement of the animals. Though these new restrictions appear to have had limited effect on the activity of most of the interviewees, one of the herders (D) has decided to enclose his animals full-time, and another (G) now does so in summer. There is, however, no question about the fact that until the recent past pigs were invariably free throughout the year. Those pigs which are free-range live on communal land and can roam as they like, though they rarely venture more than 10km away. Entire boars tend to wander off far more frequently and they are occasionally lost; indeed, they may not to be available when required, in which case a sire may need to be hired from another breeder. In winter when they are self-sufficient, pasturing in woodland, the pigs may be difficult to find, and the common practice of attracting them with food at slaughtering time may not work; in these cases the animals may be shot.

Although all free-range pigs similarly pasture freely during the day, their nocturnal behaviour is more variable: some regularly go back to a pen/sty to sleep; others do so only in winter; and others yet prefer to find shelter in caves, hollow trees and shrubs at any time of the year. Factors that may influence their decision include external temperature, the adequacy and comfort of the sty, and the availability of water sources in the countryside. Water is an important resource for pigs and, particularly in summer, they are often found pasturing along streams or near ponds, springs and lakes, which we also witnessed (Fig. 15.16). It would not be unusual for the herder not see his pigs for a month or more, particularly in winter, though one of our informers (H) checked his pigs at least twice a week. Most of the herders were, however, sceptical of our account of a Corsican breeder visiting his pigs in the mountains only two or three times a year (Albarella *et al.* 2007). In summer, when less natural food is available, the pigs need to be helped and require greater level of control. The most vulnerable animals are sows with piglets, and when they occur, they are looked after closely.

Because they mainly occupy communal land, which is not cultivated, the pigs living on the plateau are rarely in a position to cause damage to crops. This was confirmed to us by the elderly farmer, who said that he had never had problems caused by pigs, as these would physically be separated from the crop fields. Consequently iron wire inserted into the pigs' snouts, which is often used in Corsica (Albarella *et al.* 2007) and Greece (Halstead and Isaakidou, in this volume) to deter pigs from digging in crop fields, generally has no useful function in this area. Nevertheless, two of our informers did fit their pigs with iron wire, as there was a risk that the pigs could trespass cultivated fields. However, one of them (E) complained that in doing so the pig could only graze and would therefore not gain sufficient weight. Another herder specifically said that this detriment to the pig's development was the reason

Fig. 15.16 A pig enjoying a walk in the water on the Orgosolo plateau. Photo by U. Albarella.

why he did not use it, combined with the fact that in his area the cultivated fields were fenced. Other reasons were mentioned for the use of iron wire in the snout, including the prevention of animals from wandering off and of boars from killing sows and piglets.

Mobility

Work that we previously did in northern Sardinia revealed no indication of a seasonal movement of the pigs, though in Corsica some accounts of the movement of animals from the south of the island to the Castagniccia region in the north-east, where pigs would feed on sweet chestnuts, emerged (Molenat and Casabianca 1979; Albarella *et al.* 2007). Sheep, goat and cattle transhumance is still practiced in Sardinia today, and most of the herders informed us that until the recent past this was practiced for pigs too, though for different reasons and following separate routes. Up to the 1970s it was common for pigs to be driven from the highlands of our study area to plains and valleys, where they would feed on the stubble of wheat and barley after the harvest. The Campidano valley (Fig. 15.11), in the south-west of the island, was the most commonly cited lowland, but various plains around Cagliari (in the south) and Oristano (in the west) were also mentioned. This transhumance followed an opposite altitudinal gradient to the one generally practiced with grazing animals: the pigs would spend the winter in the mountains and the summer in the plains, as this, in terms of food availability, was the most efficient strategy. The distance covered by this journey would be substantial, 100km or more, and the animals would feed in the area for about three months. Up to 100 to 150 animals could be driven along this route, under the guidance of two or three herders (therefore an average of one herder for 50 pigs seems to have been required). In later years the journey was sometimes made by train. Generally the pig owners themselves would travel with the pigs to the plains, where a small charge was paid to the farmers who owned the grain fields. An alternative possibility was for the breeders to pay seasonal labourers – generally very poor people – to drive the pigs, so the former could remain in the upland villages with their families (Fig. 15.11).

In addition to the seasonal movement of pigs to the lowlands, other occasional movements of the animals occurred in years when the acorn production on the plateau had been poor. Typically the pigs would be driven from Supramonte and Ogliastra to other mountainous areas, such the Monte Limbara in the north of the island, where the acorn production might have been better (Fig. 15.17). Another, geographically closer, area where the pigs could be driven is represented by the western slopes of the Gennargentu Mountains, around the villages of Desulo and Aritzo. The main attraction of this area was the abundance of sweet chestnuts, which therefore could provide a useful alternative to acorns. In addition, pigs could be driven to urban centres, such as Cagliari, to be sold at the market (herder F).

The existence of transhumant pigs in Sardinia is mentioned by Le Lannou (2006), who, travelling across the island in the 1930s, had the opportunity to witness this movement directly and also to publish a photograph of transhumant pigs in Aritzo (Le Lannou 2006, Tav.XIIB). He does not, however, provide a reason for the movement of the pigs, limiting his comment to the fact that pigs are subject to transhumance just like sheep. His interpretation, however, that sheep are moved to the valleys to escape the cold of the winter in the mountains (Le Lannou 2006, 218) cannot be applied to pigs, which were moved around the landscape in search of better food rather than a more agreeable climate.

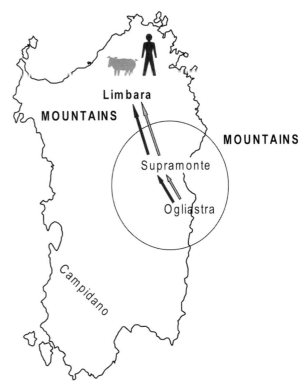

Fig. 15.17 A representation of the 'bad year' (for mast production) movement of pigs and people from the highlands of Ogliastra and Supramonte to the other highlands of Monte Limbara. Compiled by A. Trentacoste.

Other Livestock

As mentioned above, specialised pig keeping was common in the past but nowadays it is not regarded to be sufficiently profitable due to the competition of national and international markets. Therefore pig keeping must be integrated with other economic activities, of which the breeding of other livestock is the most common. With one exception all our informants kept other animal species, and even the one breeder who currently only keeps pigs mentioned that it was family tradition to breed goats as well. All but one of the interviewed pig herders kept goats and many also mentioned sheep and cattle. Writing in the 18th century Manca Dell'Arca (1780) suggested that in Sardinia pig husbandry was traditionally associated with the keeping of goats, and the evidence we gathered indicates that, to some extent, this practice has survived to these days. Goats would be kept mainly for milk and dairy products, with meat largely regarded as a by-product. Sheep and cattle would variably be used for milk and meat, whereas wool production seems to be of little interest. Due to the spread of motorised equipment cattle are no longer used for traction, but there is still vivid and widespread memory of their use for this purpose. Other animals more rarely mentioned include horses, donkeys and chickens. Although there is a rather clear separation in the area between 'farmers' and 'herders', quite a few of the pig herders were also engaged in agricultural activities such as the management of vineyards, olive groves, hay meadows and vegetable gardens.

Conclusions

Overall, the second phase of our ethnoarchaeological work in Sardinia has provided useful results in the following areas:

- The sample of interviews that we presented as part of the first phase of this project (Albarella *et al.* 2007) has been enhanced. Many confirmations to our previous observations have been obtained and the characteristics of several management practices have been further clarified;
- The geographic coverage of our ethnoarchaeological research has been extended, thus allowing us to describe previously unreported practices;
- The particularity of the central Sardinian context has been demonstrated, confirming our assumption that this was the most useful area on the island to investigate traditional practices of pig husbandry.

Many of the details about pig husbandry that we have collected as part of this project constitute an aid to archaeological interpretation similarly to those collected from other parts of Europe by Halstead and Isaakidou (in this volume) and Hadjkoumis (2010). Interpretation of various aspects of archaeological animal bone assemblages, including species frequencies, kill-off patterns, sex ratios and biometry, can greatly be enhanced by an understanding of the challenges met by contemporary herders operating in societies with a low level of industrialisation. Examples of such applications have been provided in our previous work and will not be repeated here. There are, however, a few additional points that may be worth highlighting:

1. Specialised pig herders, which nowadays no longer seem to exist, were a common occurrence in the recent past of Sardinia. Their presence suggests an economy potentially more similar to the style of management of the Spanish *dehesa* (Hadjikoumis 2010) than it may currently appear. The viability of elements of the society to make a living based entirely on pigs needs to be taken into account when interpreting ancient societies.

2. Although the Sardinian 'breed' can farrow twice a year, as pigs also did in Roman times (Varro II,IV.14), work in central Sardinia has revealed that even in recent years the traditional regional type, due to shortage of summer food, would commonly give birth only once a year. Manca dell'Arca (1780, 324) also writes that the ability to produce two litters per year was dependent on sufficient nourishment. Wild sows can also potentially farrow twice in a year, though the necessary conditions rarely occur (Toschi 1965). It is likely that in the distant past, when industrially produced food was unavailable and the highly privileged conditions described by the classical authors were rare, single farrowing was the norm for unimproved pigs. This has important implications for archaeological interpretation of seasonal patterns, though the variable birth season characteristic of the Mediterranean area is an issue that cannot be ignored.

3. Like in Greece (Halstead and Isaakidou, in this volume) and Corsica, as well as in contemporary and early modern Sardinia (Della Marmora 1839) and Roman Italy (MacKinnon 2001), a double system of free-range and enclosed pigs is in place in central Sardinia; however, an extensive style of management is definitely predominant. Widespread communal land makes this possible. The potential clash between free range livestock and crop field can thus be avoided not only through a device that will stop pigs from burrowing, but also through landscape management. This has important implications for archaeological interpretation.

The most important new element that we have identified as part of this new project concerns patterns of human and animal mobility associated with livestock management. The double – lowland and highland – style of settlement, with the periodic movement of people from one to the other, has highlighted the occurrence of a relatively simple model of resource exploitation (Fig. 15.3), which takes advantage of different environments and food sources. The splitting of the community into separate settlements fulfilling different functions represents a scenario that we may find replicated in the archaeological record. In addition, we should consider that many of the animals living on the plateau were eventually consumed in the main village in the valley, which means that this is where their remains are expected to be found. This represents a further reminder of the important distinction between production and consumption that we should make in the interpretation of the zooarchaeological evidence.

The economic and mobility model in place in central Sardinia is made possible by the existence of vast areas of communal land in the highlands. Such communality was a common occurrence in Sardinia until the notorious Editto delle Chiudende (Law of Enclosures) was imposed in 1820 by the Piemontese government (Mientjes *et al.* 2002). This law gave landowners the right to enclose their land and abolished the system of common rights in the use of the land. In the traditionally pastoral area of central Sardinia this law was met with fierce opposition from shepherds who, through the implementation of this legislation, had lost their free access to pasture. It is therefore probably not accidental that, in this area, both extensive common land and free-range livestock management still survive. Before 1820, this extensive animal management system was probably spread across the island.

Another important element that has emerged from our work in central Sardinia is represented by the evidence – provided by several informers – of a pig transhumance, which was integral part of the local system of pig management and is a poorly documented practice in the history of husbandry. Pigs were moved over substantial distances, even exceeding 100 km, which is not entirely surprising if we consider the remarkable agility of the Sardinian 'breed'. These animals are very adept to walk and even run (Fig. 15.18) as part of their survival strategy in an environment that is, for most of the year, fairly harsh. Such pig types were certainly common in the past and should alert us to the fact that the presence of pig bones in archaeological sites must not necessary be taken as evidence of sedentism. Unquestionably pigs are not as gregarious as cattle or sheep, but the information we gathered from our interviewees indicates that they can be driven, with, on average, one herder able to take care of about 50 pigs.

Long distance movement of pigs could have a number of different motivations. Of these the most obvious and best documented is the search in winter of suitable woodland, where pigs would feed on mast. Our study area includes abundant woodland and therefore the pigs would not need to be moved unless a specific year had been bad for acorn production. In these cases the pigs were indeed driven away towards areas that had different types of mast or had experienced more successful woodland production. Le Lannou (2006) – writing in the 1930s – regarded the villages of Aritzo and Desulo as typical destinations of pig transhumance. Quite possibly many of these pigs reached those areas as part of regular seasonal movements from regions that were not equally rich in woodland. Similar reasons have been advocated for cases of pig transhumance in 16th and 17th century Spain (Moraza 2005) and France (Moriceau 2005), where pigs would spend two months or more in woodland areas. In France the practice seems to go back to medieval times at least.

Our work has, however, highlighted the occurrence of an additional type of transhumance, occurring in summer rather than autumn or winter. As part of this seasonal practice the pigs would be moved from the mountains to the plains, where they could feed on cereal stubble. This was the most common long mobility pattern in the mountains of central-eastern Sardinia, where the challenging time of the year for pig feeding is clearly the summer rather than the winter.

Finally, pigs could be moved for market rather than pasturing reasons, though it is conceivable that the two activities could be combined. For instance, pigs that were driven from *Ogliastra* to *Campidano* in summer may then have made the shorter journey to the southern urban centre of Cagliari (the largest city in Sardinia) where they could be sold at the market. Instances of pigs being moved all the way from central-eastern Sardinia to Cagliari were, however, also mentioned. Long distance movement of pigs for market purposes has been suggested by Steele (1981) for Roman Italy on the basis of zooarchaeological evidence.

With so much movement of animals from one area to the other, it is therefore not surprising that regional pig types do not seem to occur and that a generic Sardinian 'breed' is – or at least was – spread all over the island. Interbreeding of different populations must have been the norm throughout history. The implications of these husbandry strategies for the organization of human societies are substantial. For instance men and women had different patterns of mobility with the latter clearly being far more sedentary, as herding was clearly a man's activity. Status is also clearly entangled with livestock management; we have, for instance, seen that often the summer pig transhumance was carried out by poor seasonal workers. What is also important for archaeological interpretation is that both people and animals may have left traces of their activities – and in some cases their remains – in places that are different from those where they were born or spent most of the year. The other useful consideration concerns trade and cultural contact between different sub-regions, which clearly must have been substantial.

Pigs are not, and were not, the most numerous livestock in Sardinia; historically they have been substantially outnumbered by sheep (*cf.* Le Lannou 2006; Mientjes 2008). Yet they have played a very important role in the shaping of Sardinian society, and although specialized pig breeding no longer exists, pig husbandry remains a very important economic resource, particularly in Ogliastra. Plainly, Sardinian settlement and human mobility patterns have been much influenced by pig management. This has very important implications for archaeological interpretation in Sardinia itself, but also elsewhere. Much of the evidence that we have collected through our interviews with the Sardinian herders is affected by local climatic, environmental and cultural factors. This is why we must beware of applying our various mobility models (Figs. 15.3, 15.11 and 15.16) *tout court* to archaeological contexts, particularly outside Sardinia. However, the

Fig. 15.18 A pig in full gallop on the Orgosolo plateau. Photo by U. Albarella.

various individual elements of our reconstructions represent useful food for thought for archaeological interpretation and highlight the cultural value and importance of practices that are today seriously endangered. These traditions represent an important element of our heritage, whose preservation should be of global and not merely local concern.

Acknowledgements

The fieldwork phase of this project was generously funded by a British Academy grant awarded to U. A. in 2005. U. A. would also like to thank the Royal Society, which, in 2006, funded his trip to the ICAZ conference in Mexico City, where the oral version of this paper was first presented. We owe a huge debt of gratitude to many local people who helped us in the summer of 2005, Nazarino and Caterina Patteri (Orgosolo), Rosina Piras and Paola Lobina (Villanova Strisaili) and Filippo Sotgia (Dorgali) in particular. They generously made their knowledge available, and in many cases also opened their houses, precious hams and bottles of wine to their inquisitive visitors. We are especially indebted to Lino Ruiu, whose Hotel Santelene in Dorgali, was our 'base camp' for the duration of the project. Apart from his hospitality, it was Lino's enthusiasm and support that contributed substantially to the success of the project. Many of the issues discussed in this paper have been the subject of stimulating conversations with Angelos Hadjikoumis, Paul Halstead and Jean-Denis Vigne, to whom we are also grateful. The title of this paper is inspired by the novel 'A Year on the Plateau' by the Sardinian writer Emilio Lussu.

References

Albarella, U. (2006) Pig husbandry and pork consumption in medieval England. In C. Woolgar, D. Serjeantson and T. Waldron (eds.) *Food in Medieval England: diet and nutrition*, 72–87. Oxford, Oxford University Press.

Albarella, U., Manconi, F., Rowley-Conwy, P. and Vigne, J.-D. (2006) Pigs of Sardinia and Corsica: a biometrical re-evaluation of their status and history. In U. Tecchiati and B. Sala (eds.) *Archaeozoological Studies in Honour of Alfredo Riedel*, 285–302. Bolzano, Province of Bolzano.

Albarella, U., Manconi, F., Vigne, J.-D. and Rowley-Conwy, P. (2007) The ethnoarchaeology of traditional pig husbandry in Sardinia and Corsica. In U. Albarella, K. Dobney, A. Ervynck and P. Rowley-Conwy (eds.) *Pigs and Humans: 10,000 years of interaction*, 285–307. Oxford, Oxford University Press.

Albarella, U., Dobney, K. and Rowley-Conwy, P. (2009) Size and shape of the Eurasian wild boar (*Sus scrofa*), with a view to the reconstruction of its Holocene history. *Environmental Archaeology* 14(2), 103–136.

Cetti, F. (1774) *Quadrupedi, Uccelli, Anfibi e Pesci di Sardegna*. Anastati reprint. Cagliari, GIA Editrice.

Columella. *L'Arte dell'Agricoltura*. (1997) Translated from Latin by R. Calzecchi Onesti. Torino, Einaudi.

Della Marmora, A. (1839) *Viaggio in Sardegna. La geografia fisica e umana. Vol. I*. Nuoro, Editrice Archivio Fotografico Sardo.

Hadjikoumis, A. (2010) *The origins and evolution of pig domestication in Spain*. Unpublished PhD dissertation, University of Sheffield.

Larson, G., Albarella, U., Dobney, K., Rowley-Conwy, P., Schibler, J., Tresset, A., Vigne, J.-D., Edwards, C., Schlumbaum, A., Dinu, A., Balasescu, A., Dolman, G., Tagliacozzo, A., Manaseryan, N., Miracle, P., Van Wijngaarden-Bakker, L., Masseti, M., Bradley, D. and Cooper, A. (2007) Ancient DNA, pig domestication, and the spread of the Neolithic into Europe. *Proceedings of the National Academy of Sciences of the United States of America* 104(39), 15276–15281.

Le Lannou, M. (2006) *Pastori e contadini della Sardegna*. Translated from French by M. Brigaglia. Cagliari, Della Torre (Originally published in 1941).

MacKinnon, M. (2001) High on the Hog: linking zooarchaeological, literary, ad artistic data for pig breeds in Roman Italy. *American Journal of Archaeology* 105, 649–673.

Manca Dell'Arca, A. (1780) *Agricoltura di Sardegna*. Napoli, Vincenzo Orsino.

Mientjes, A. (2008) *Paesaggi Pastorali. Studio etnoarcheologico sul pastoralismo in Sardegna*. Cagliari, CUEC.

Mientjes, A., Pluciennik, M. and Giannitrapani, E. (2002) Archaeologies of recent rural Sicily and Sardinia: a comparative approach. *Journal of Mediterranean Archaeology* 15(2), 139–166.

Molenat, M. and Casabianca, F. (1979) *Contribution à la maîtrise de l'elevage porcinextensif en Corse*. Bulletin Technique du Departement de Genetique Animale 32. Jouy-en-Josas, Institut National de la Recherche Agronomique.

Moraza, B. A. (2005) La transhumancia desde el sistema ibérico al Pirineo occidental: el pastoreo de ganado porcino entre la sierra de Cameros (Soria-La Rioja) y el País Vasco a fines de la Edad Media. In A. Catafau (ed.) *Les ressources naturelles des Pyrénées du moyen âge à l'époque moderne. Explotation, gestion appropiation. Actes du congrès international Resopyr 1*, 221–238. Perpignan, Presses Universitaires de Perpignan.

Moriceau, J. M. (2005) *Histoire et géographie de l'élevage français, du Moyen Âge à la Révolution*. Paris, Fayard.

Onida, P., Garau, G. and Cossu, S. (1995) Damages caused to crops by wild boars (*S.scrofa meridionalis*) in Sardinia (Italy). *Ibex* 3, 230–235.

Porter, V. (1993) *Pigs. A Handbook of the Breeds of the World*. Mountfield, Helm.

Rappaport, R. A. (1968) *Pigs for the Ancestors. Ritual in the ecology of a New Guinea people*. New Haven and London, Yale University Press.

Redding, R. and Rosenberg, M. (1998) Ancestral pigs: a New (Guinea) model for pig domestication in the Middle East. In S. Nelson (ed.) *Ancestors for the Pigs: pigs in prehistory*, 65–76. MASCA Research Papers in Science and Archaeology 15. Philadelphia, University of Pennsylvania.

Rubel, P. and Rosman, A. (1978) *Your Own Pigs You May Not Eat. A Comparative Study of New Guinea Societies*. Canberra, Australian National University Press.

Scrivener, P. (2010) *The Ethnography of Pig Husbandry: a review of traditional methods of relevance to archaeology with original data from the New Forest*. Unpublished BSc dissertation, University of Sheffield.

Sillitoe, P. (2003) *Managing animals in New Guinea. Preying the game in the Highlands*. London and New York, Routledge.

Steele, D. (1981) The analysis of animal remains from two late Roman middens at San Giovanni di Ruoti. In M. Gualtieri, M. Salvatore and A. Small (eds.) *Lo scavo di S.Giovanni di Ruoti e il periodo tardo-antico in Basilicata*, 75–84. Bari, Adriatica.

Toschi, A. (1965) *Fauna d'Italia. Mammalia. Lagomorpha – Rodentia – Carnivora – Artiodactyla – Cetacea*. Bologna, Calderini.

Varro, M. T. (1993) *On Agriculture*. Translated from Latin by W. D. Hooper and H. B. Ash. LOEB Classical Library 283. Cambridge, Massachusets and London, England, Harvard University Press,

Wiseman, J. (2000) *The pig: a British history*. London, Duckbacks.

16. A pig fed by hand is worth two in the bush: Ethnoarchaeology of pig husbandry in Greece and its archaeological implications

Paul Halstead and Valasia Isaakidou

'Traditional' pig husbandry in Greece is discussed, drawing on personal observations and interviews with retired herders. Informants managed mainly unimproved pigs (with some wild or improved admixture) on a range of scales (from specialist breeding to fattening a pig for domestic consumption) and in various ecological settings (highland/lowland; woodland/cultivated; deciduous/evergreen). Fattening of a single pig, to provide lard and preserved meat for household consumption and hospitality, was widespread. Large herds were few, reflecting the pig's lack of secondary products and the ease with which a few sows with large litters could satisfy a restricted market for meat. Pig herders were normally part-time and temporary specialists. Although husbandry practices varied, household pigs were more intensively fed, gained weight faster, achieved larger carcasses, put on more fat, and produced larger litters than those raised in herds. Traditional pig husbandry and consumption in Greece has much in common with that in the west Mediterranean, but presents some significant contrasts with highland New Guinea.

Implications for past pig husbandry are briefly considered. The difficulty of isolating herded pigs from wild boar, and the potentially negative consequences of failing to do so, raises questions for palaeogenetic studies of pig domestication, but may shed light on the scale and nature of early pig husbandry. Moreover, while some intensive fattening is implied textually for Bronze Age Greece and Roman Italy, Neolithic exploitation perhaps had less in common with recent Greek fattening of a pig for domestic consumption than with highland New Guinea rearing for collective feasting.

Key words: pig, Greece, intensive, extensive, ethnoarchaeology

Introduction

All archaeological interpretation is informed by observations in the present, and 'ethnoarchaeological' studies of animal exploitation (*e.g.* Binford 1978) have played a significant role in promoting awareness of this principle. Zooarchaeologists active in southern Europe and southwest Asia have access to a wealth of *local* and thus ecologically relevant studies, ethnographic and ethnoarchaeological, on 'traditional' management of sheep and goats, albeit with a bias towards large-scale, specialised herding rather than small-scale, mixed husbandry. Such information is sparse, however, for pigs (Albarella *et al.* 2007). The first section of this chapter presents original information on 'traditional' pig husbandry and consumption in Greece, emphasising contrasts between large- and small-scale herders. The second section highlights similarities and contrasts with two regions in which pig husbandry is of considerable cultural significance: the West Mediterranean and the socially and ecologically very different highlands of New Guinea. The final section briefly explores implications for domestication, management and consumption of pigs in the distant past in Europe and especially Greece.

'Traditional' Pig Husbandry in Greece

Nature of the Evidence

During zooarchaeological work in Greece, we have encountered current and retired pig-farmers, with experience that ranges from breeding substantial herds for market to fattening one or two 'household' pigs for domestic consumption. What follows is based partly on personal observation, especially of extensively managed herds, and partly on oral accounts, particularly of consumption of pigs before refrigeration became widely available. Information on extensively managed herds, mainly in northern Greece

No.	Location	Altitude	Vegetation	Herd control	Breed	Date
1	Ag Dimitrios, Mt Olimbos	c. 1000m	deciduous oak-beech woodland	enclosed	farmed wild boar	current
2	Exokhi, Pieria hills	c. 200m	fields + deciduous woodland	herded	traditional (x wild boar)	1950s–70s
	Exokhi, Pieria hills	c. 200m	deciduous oak-hornbeam woodland	enclosed	traditional x wild boar	current
3	Ano Milia, Pieria Mts	c. 1000m	deciduous woodland	free-range	traditional + farmed wild boar	current
4	Paliambela-Kolindrou, lowland C Macedonia	c. 60m	stubble fields	herded	traditional	1930s–
5	Assiros, lowland C Macedonia	c. 200m	stubble fields, orchards	herded	traditional	1950s–60s
6	Megali Panagia, Khalkidiki peninsula	c. 400m	fields + deciduous/evergreen oak woodland + beech/deciduous oak woodland	herded	traditional	1970s–
7	Varvara, Khalkidiki peninsula	c. 600m	fields + deciduous/evergreen oak woodland + beech/deciduous oak woodland	herded	traditional x farmed wild boar	latter 20 c. + current
	Varvara, Khalkidiki peninsula	c. 600m	beech + deciduous oak + mixed evergreen woodland	free-range	traditional x farmed wild boar	current
8	Olimpiada, Khalkidiki peninsula	c. 0–200m	fields	herded	traditional	1960s–70s
9	Ierissos, Khalkidiki peninsula	c. 300m	fields + mixed evergreen-deciduous oak woodland	free-range	traditional x wild boar	current
10	Asoutaina, hills of Messinia	c. 800m	fields + mostly evergreen oak woodland/scrub	free-range (Sept-Dec)	traditional	1950s–70s
	Asoutaina, hills of Messinia	c. 800m	abandoned fields + mostly evergreen oak woodland/scrub	free-range	traditional x wild boar	1990s–
11	Palaio Loutro, hills of Messinia	c. 500m	fields + mostly evergreen woodland	herded	traditional	mid–20 c.
12	Aloides, lowland C Crete	c. 400m	mostly evergreen scrub	free-range	traditional	1930s–40s
13	Yeni Gave, lowland C Crete	c. 200m	evergreen scrub with carobs and deciduous oaks	free-range	traditional	1940s–

Table 16.1 Location and context of extensively managed pig herds discussed

but also in the southern mainland and island of Crete (Fig. 16.1), is drawn from a range of husbandry regimes and ecological contexts (Table 16.1). Information on fattening and consumption of household pigs and on small-scale pig breeding, in several regions (Fig. 16.1), is mostly from lowland mixed-farming communities, but also from upland mixed-farming settlements (Anogia, Kipourio and Vitsa at 800–1000m) and summer villages of transhumant pastoralists (Aetomilitsa and Fourka at 1500m). Ecological context is less important for pigs largely reared near home than for those herded extensively, but the distinction between these scales of husbandry is fluid and rather arbitrary, and contextual details are given below where relevant.

Most of the herds considered were based on 'traditional' breeds (Table 16.1) of long-legged and long-snouted pigs with dark or piebald hairy skins (Figs. 16.2 and 16.3). In Pieria, such animals are called 'bristly pigs' (*gatzogoúrouna*), but are often referred to as 'wild' (*ágria*) to distinguish them from improved (*ímera*, literally 'tame') animals. We targetted informants who reared unimproved pigs, the dietary needs and productivity of which might be relevant to the past. To varying degrees, herds of traditional breed have interbred with wild boar or farmed pigs of wild stock, yielding some animals of uniformly dark and bristly appearance with larger tusks (Figs. 16.4–16.6). The woods of the Khalkidiki peninsula shelter animals that – apart from their tameness and perhaps a bell round the neck – so

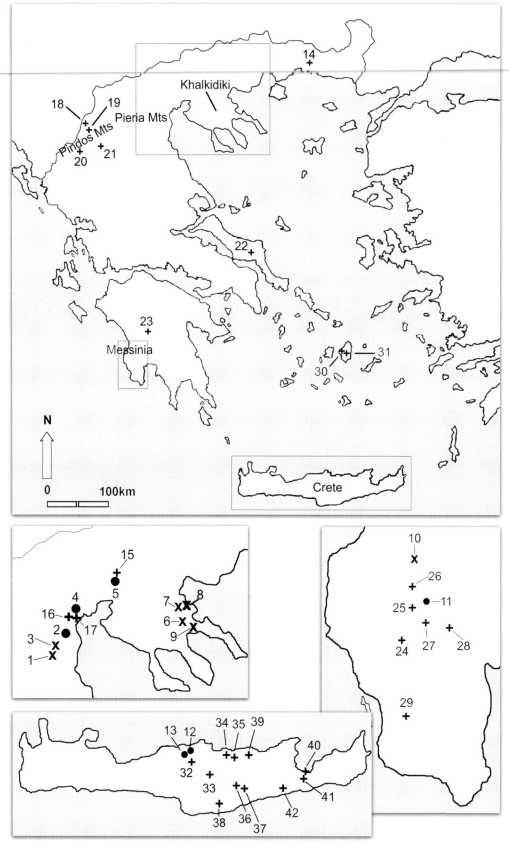

Fig. 16.1 Map of Greece showing location of recent 'extensive' (nos. 1–13) and 'household' (2, 4, 5, 11–13) pig rearing discussed in this chapter. Key: 1 Mt Olympus, 2 Exokhi, 3 Ano Milia, 4 Paliambela-Kolindrou, 5Assiros, 6 Megali Panagia, 7 Varvara, 8 Olimpiada, 9 Ierissos, 10 Asoutaina, 11 Palaio Loutro, 12 Aloides, 13 Yeni Gave, 14 Neo Sidirokhori, 15 Mavrorakhi, 16 Kolindros, 17 Aiginio, 18 Aetomilitsa, 19 Fourka, 20 Vitsa, 21 Kipourio, 22 Tharounia, 23 Karitaina, 24 Khora (Messinia), 25 Metaxada, 26 Potamia, 27 Maniaki, 28 Milioti, 29 Kinigou, 30 Khora (Naxos), 31 Filoti, 32 Anogia, 33 Ano Asites, 34 Knossos, 35 Skalani, 36 Partira, 37 Ag. Semni, 38 Dionisi, 39 Kalo Khorio Pediadas, 40 Pakhia Ammos, 41 Vasiliki, 42 Mirtos

Fig. 16.2 Herded pigs of 'traditional' breed in deciduous woodland, near Megali Panagia, Khalkidiki peninsula

Fig. 16.4 Adult free-range pigs of 'wild' type, 'hoovering up' grain scattered to attract them, near Varvara, Khalkidiki peninsula

Fig. 16.3 Free-range pigs of 'traditional' breed in deciduous woodland, 'hoovering up' grain scattered to attract them, near Varvara, Khalkidiki peninsula

Fig. 16.5 Young free-range pig of 'wild' type, near Varvara, Khalkidiki peninsula

Fig. 16.6 Piglet of 'wild' type, near Varvara, Khalkidiki peninsula

resemble wild boar that some are mistakenly shot by hunters from Athens. Some herders have introduced wild stock (usually breeding males), mostly from farms in Greece and abroad but also young pigs trapped by hunters, to exploit a growing market for lean and 'gamey' meat. In addition, wild males break down fences to mate with domestic females and the latter may seek out wild males when in season. Unplanned interbreeding has probably increased in recent years, as declining cultivation in agriculturally marginal areas has enabled domestic pigs to roam freely without damaging crops. Some herders try to minimise interbreeding with wild boar (*e.g.* by shutting up domestic sows in season), because the offspring put on weight slowly or may disappear into the woods and become feral.

Most of the household pigs described by elderly informants were of traditional breed, but some recent examples were improved animals or wild boar crosses kept by 'hobby farmers'.

Herd Control

Like domestic ruminants, most herds of pigs used to forage around the landscape under supervision of a herdsman,

Fig. 16.7 Herded pigs of traditional breed on cereal stubble, near Megali Panagia, Khalkidiki peninsula; at other times of year, these pigs forage free-range in woodland

Fig. 16.9 Free-range pig of 'traditional' breed on abandoned fields with scattered fruit and nut trees, Asoutaina, Messinia

Fig. 16.8 Free-range pigs of 'traditional' breed crossed with wild boar in deciduous woodland, near Varvara, Khalkidiki peninsula

responsible primarily for keeping them out of growing crops (Fig. 16.7). Dogs often guarded the herd, but Stavros in Khalkidiki believed familiarity with dogs undermined his pigs' instinct to chase away wolves. During the hottest summer months, some herders took their pigs out at night. Pigs foraging with a herdsman might walk several kilometres in a day. We have not encountered transhumant *herds*, but some Vlach informants took a few pigs with their sheep flocks on the annual migration between Aetomilitsa or Fourka in the Pindos mountains and winter villages 150 and 130km distant as the crow flies. In 1922, one refugee from European Turkey walked more than 400km with 50 sheep and a pig to Aiginio in Pieria.

Some supervised herds ranged freely at times of year when they foraged in woodland or scrub: for example, at Aloides on Crete, where there are no large four-legged predators, but also in Khalkidiki, where Stavros' pigs chased away wolves. Herds of pigs range freely all year in the hills of Khalkidiki (Fig. 16.8) and Messinia (Fig. 16.9), attracted home each evening by fodder – in modest quantities sufficient to entice rather than sustain (Figs. 16.3 and 16.4). These animals (sows and young) were, despite admixture of wild boar stock, unconcerned at unfamiliar humans wandering among them. Herders of farmed boar consider adult males always potentially dangerous, but adult females aggressive only at mating and farrowing. The contraction of cultivation, that has allowed free-range herds to expand, has also resulted in crop raiding by growing numbers of wild boar and other game being focussed on, and often ruining, the few remaining crops in areas such as the Pindos Mountains.

Households fattening one or two pigs often kept them in the yard and kitchen garden, perhaps tied to a post (or even – in one account from Assiros – to the dog). In Paliambela-Kolindrou, household pigs were confined to a small makeshift pen. Alternatively, a family member might take the pig(s) to graze with the household's cattle (Assiros) or sheep and goats (Kinigou), or a hired herder might mind the household pigs of a whole village (mid-twentieth century Assiros). In Assiros, a wandering household pig (or sheep or goose) was liable to be eaten by a neighbour. To avoid damage from rooting, pigs at Filoti on Naxos were shut up in winter, while wandering pigs in the hill villages of nineteenth century central Greece wore angular collars that stopped them getting through gaps in fences (Loukopoulos 1983, 385–386). Pigs in *herds* at Paliambela-Kolindrou likewise had their snouts wired to prevent digging. Household pigs allowed to forage in the garden or taken out to graze were often confined for the last month or two before slaughter – in a hut with slatted floor at Assiros – to aid fattening. Stall-fed pigs, in particular, could become very tame: Maria recalls the family pig in Partira rolling on its back to be petted before feeding.

Reproduction

Sows bred from one or two years of age, depending on how well fed they were. Female piglets might be selected for breeding on visual evaluation (*e.g.* those with most teats) or quality of mother. Some herders retained sows as long as they produced enough milk to suckle (to 7–8, occasionally 10, years of age). Others slaughtered sows at 4–5 years, before the meat got too tough for anything but sausages or 'because older females might wander off with their piglets'. Some households, that reared pigs primarily for domestic fattening and sold a few surplus piglets, bred from even younger sows. For example, in Assiros, Tasoula's father would keep his sow for one or two litters and then slaughter her, to be replaced by one of her offspring. Recently, in Messinian Khora, Stavroula was fattening a male and a female piglet, which mated. She slaughtered the male as planned at one year, but kept the female to two years, by which time she had produced two litters. Stavroula never rears a pig beyond two years of age.

Male pigs can breed from their first year, as Stavroulla's unplanned litter illustrates. Breeding males may be killed fairly young (at 2–3 years), even in large herds, because the weight of older males can impede mating. On a household scale, they may be killed even younger. For example, in Metaxada, Grigoris' two sows produced piglets, most of which he sold off at a few weeks of age. He fattened one or two piglets for household consumption and kept one entire male to mate with his and (for a token fee) his neighbours' sows. The male was castrated at one to two years old and slaughtered a month later, producing a lean carcass that probably discouraged many households from keeping a breeding male. For his herd of *gatzogoúrouna* pigs at Exokhi, Yorgos reckoned on one boar for ten breeding sows (1:15 if the boar was very well fed), but keeping more than one breeding male may be difficult. Males that have grown up together are considered less antagonistic, although one breeding male at the Mt. Olympus boar farm recently used his tusk to castrate his twin brother while the latter was engaged in reproductive duties.

The length of gestation in pigs is memorably described as 'three months, three weeks and three days'. Extensively managed sows often disappear before farrowing, to build a nest of branches and mud like their wild counterparts, returning with piglets a week or so later. Sows reared together sometimes share a nest. A hidden sow, with snout protruding from the nest, presents a formidable deterrent to wolves, but the boar farm on Mt. Olympus loses piglets to foxes and martens, which escape by their speed and tree-climbing ability respectively, and a bear recently killed a defending sow.

The size of litters is very variable, with reported ranges of 2–13 from the boar farm on Mt. Olympus, of 5–10 and 5–15 from free-range herds in Khalkidiki, and of 8–15 from the free-range herd at Asoutaina. In the latter herd, which received some unplanned influx of wild genes, sows of 'domestic' appearance are said to produce more piglets than their 'wild' hairier counterparts. Young and old sows produce smaller litters than prime adults: for example, Yorgos' sows of traditional breed herded around Exokhi produced four piglets in their first litter and 6–7 thereafter, while the owner of a similar herd near Varvara expected up to 4–5 piglets from yearling sows and larger litters thereafter. Well-fed yearlings, however, may have large litters: for example, at Khora on Naxos, Manolis regularly bought and fattened the odd piglet, but one year kept two females, each of which produced a litter of twelve.

In Messinian Khora, the two litters of Stavroula's young sow added up to 18 piglets. Litter size also depends on the condition of the breeding males: when recent expansion of the herd at Exokhi outstripped the males' stamina, the first sow mounted by each boar produced 6–8 piglets, but those mated subsequently only 3–4. The effects of inbreeding may be severe and rapid: at Assiros in the 1950s, Vasilis built up a herd of 200 pigs from a single sow, using her piglets as breeding males, but after a few years litter sizes fell from 10–11 to 3–4 and congenital defects increased until he introduced new males. A realistic average for an extensively managed but well-fed adult sow, with good access to a healthy boar, may be 6–8 piglets, but inadequate diet, overworked males and inbreeding lower this figure. Conversely, well-fed household pigs, even yearlings, may produce significantly bigger litters. Infant mortality is also variable: early and late litters are vulnerable to cold weather; adult male pigs (as well as other predators) may kill young piglets; and the Exokhi herd suffered hugely increased losses to disease when supervised foraging around the village and temporary stalling, that shifted twice a year, gave way to enclosure and permanent housing.

Pigs are said to farrow up to five times in two years, but this is difficult in practice. Herders reckon on sows coming into season 10–15 days after they cease (or reduce) suckling. Allowing almost four months (114 days) for gestation, five litters in two years would leave 20 days of suckling per farrowing, while two litters per year could each be suckled for 55–60 days. Today, by feeding cereals to sows and offspring, most herders wean piglets from about 40 days old to achieve two or slightly more farrowings per year. A few decades ago, with less easy access to grain fodder, Yorgos in Exokhi weaned piglets from 45 days, while two herders in Paliambela-Kolindrou delayed until 50–60 days and 2–3 months to avoid stressing the piglets. Some households that maintained a single sow also report suckling of piglets until three months of age. Elderly villagers (*e.g.* from Milioti and Potamia) regard two litters per year as common for well-fed household pigs, though Apostolis' improved but underfed sow in Assiros only farrowed once a year. Two litters per year were far from the norm (as opposed to ideal) in larger herds, until cheap feed grains enabled early weaning. In Asoutaina, Mimis believes his free-range sows would not produce two litters per year without maize fodder: many now farrow around February and again in late summer, when – in dry years – they have lost weight for lack of fresh plants. Further north and at higher altitude, the farmed boar on Mt. Olympus tend to farrow in May–June, and perhaps again in winter if the earlier litter is weaned quickly (at 45–50 days, with the aid of heated farrowing pens). The second litters used to suffer very heavy losses due to cold weather, however, and the herder believes that, without sheltered accommodation and feeding, his sows would suckle for 8–11 weeks and farrow once a year, like local wild boar.

By shutting up breeding males, some herders manipulate farrowing dates to avoid extreme temperatures (winter cold and summer heat) or scarce food, to ensure piglets reach a marketable weight for Christmas, or to avoid adults foraging in ripening crops. Where breeding males are left with sows, some births occur in all seasons, but most in late winter (southern Greece) or spring (northern Greece), perhaps with a second litter in late summer or autumn. The late winter/spring litters are born to sows that mated in autumn/early winter, when they were in good condition thanks to seasonally abundant acorns and the like. For herders without access to woodland, the season of abundance may be after harvest, when pigs find fallen ears in stubble fields, and one such herder from Paliambela-Kolindrou recalls that his main farrowing period was autumn. Where just one or two sows were kept around the household or in gardens, food supply may have been less seasonal and anecdotal information suggests that such animals could farrow more or less at any time of year.

Diet

The diet of extensively managed pigs in Greece is highly seasonal in composition, quality and quantity. For pigs with access to woodland, autumn and early winter are a time of abundance: acorns of evergreen and deciduous oaks and carobs in lowland central Crete; evergreen and deciduous acorns in the hills of Messinia; acorns of deciduous oaks, with chestnuts or beechmast at higher altitude, in inland Pieria; and evergreen and deciduous acorns, beechmast and chestnuts in the coastal hills of Khalkidiki. Chestnuts, beechmast and acorns of deciduous oaks tend to fall first, followed by acorns of evergreen holm oak (*Quercus ilex*) and finally prickly oak (*Q. coccifera*). Nut fall may thus last two or even three months, between October/November and December/January. Pigs may rub against trees to shake acorns down and, conversely, delay consumption of bitter types until germination makes them sweeter. It is widely held that acorns of evergreen oaks spoil within a few weeks, whereas those of deciduous trees may be preserved for several months, kept moist under fallen leaves; in years of adequate rainfall, pigs may feed on such acorns until June or even July. Grubs consume many acorns and appeal greatly to pigs. As one herder in Varvara put it, 'the pig is always looking for delicacies' and, after eating a few acorns, sifts through the leaves looking for grubs. Autumn and winter are not always favourable: acorns tend to be abundant in alternate years, and low rainfall in an 'on year' may lead to three lean harvests in succession; beechmast may be empty in dry years. Alternative winter food includes young herbs (*e.g.* on ploughed fields), fungi and snails, but when food is scarce above ground and the earth soft pigs dig for rhizomes of ferns, grasses such as *Cynodon dactylon*, and sedges. Yorgos in Exokhi was impressed by the ability of his pigs to locate, presumably by smell, sedge roots in fields covered with snow.

In spring, pigs browse on young leaves, acorns (and grubs) may still be available in deciduous litter, and the ground is soft enough to dig for roots and 'worms'. In summer, pigs were (and occasionally still are) let into stubble fields to eat fallen ears and emerging weeds. A little later, they might be taken to threshing areas to eat crop-processing waste and then to field margins and uncultivated land for blackberries, *Arbutus unedo* fruits and fallen wild pears or figs. Especially in northern Greece, where summer rainfall is not unusual, summer annuals may flourish in harvested fields and gardens, and pigs (like sheep) enthusiastically consume purslane (*Portulaca oleracea*). In dry summers, however, food is scarce once the stubble fields have been gleaned and, apart from margins of streams, the ground may be too hard to dig. This is often a lean time, when many herders (especially if hoping for a second farrowing in autumn) now provide grain to otherwise largely self-sufficient pigs. Pigs are, literally, omnivores and 'eat even dead chickens and faeces'. They are also gourmet consumers, however, and mainly dig for food when palatable alternatives are not available above ground.

To some extent, the diet of household pigs may exhibit a similar seasonal rhythm to that of extensively managed herds. At Maniaki, household pigs were tethered in fields to eat the acorns falling from privately owned oaks. Elsewhere (Assiros, Paliambela-Kolindrou, Tharounia, Milioti, Palaio Loutro, Potamia), acorns were collected and, perhaps after beating off the cups, fed to pigs whole (they were sometimes broken for cattle, sheep and goats). In Aloides, acorns were beaten from trees with a long pole in the same way as olives. In Assiros, poor villagers without livestock gathered acorns and sold them to owners of pigs or sheep. In Potamia in the 1930s, a wet year with abundant acorns allowed pigs to be fattened for market and brought money into this marginal hamlet. A family might gather a few kilogrammes to a few sacks of acorns (and figs, wild pears, etc.) and so significantly improve the diet and body weight of one or two stall-fed pigs, but larger numbers were taken to the woods to forage.

In spring, household pigs might be tethered in the kitchen garden or fed young plants collected during weeding and, in this respect, their diet again resembled that of extensively managed herds. Through spring and early summer, household pigs were often fed the more or less watery by-product of domestic cheese making and many specialist herders of sheep (*e.g.* at Vitsa in the Pindos, Filoti on Naxos) reared a handful of pigs on whey. A large-scale cheese maker in Assiros maintained a small herd on dairy by-products and whatever they could dig up along the stream, but whey more commonly played a significant role in feeding small numbers of pigs. For similar reasons of scale, spoiled or surplus produce from gardens and orchards in summer and kitchen scraps year-round made a more significant contribution to the diet of household pigs than of larger herds. The feeding of kitchen scraps to the pig is recalled in a rhyming couplet from Kalo Khorio in Crete: 'tin kouzoulí sou kefalín tha tin ekámw goúrna, na váno t' apoplímata na trói i gouroúna' (I'll make your stupid head into a trough, to put the kitchen scraps in for the sow to eat).

Herds of pigs in the recent past were fed (at best) very modest amounts of grain when other food was scarce (dry summers, cold winters), when good diet was most critical (to sows at farrowing, piglets at weaning) or to encourage free-range animals to return home. By contrast, a household pig is often said to eat as much as an adult human. Villagers fed the household pig whatever they could manage, knowing that the less it ate, the less fat it would produce. As Fotis in Assiros observed, however, only those owning a dairy flock/herd or enough land to be full-time farmers could rear a household pig. His family (and others) had too few fields for their own grain requirements, let alone for a pig. They gave the whey from their milk cow to a relative who reared a pig (and gave them pork at Christmas). Maize (usually ground) was widely regarded as the best grain for fattening pigs, but ground barley was often given, rye too in Khalkidiki, and cultivated lupins in Messinia. Ground grains were often supplemented or replaced by cereal bran. In Messinia, pigs were commonly fattened on olive pressings (also Loukopoulos 1983, 385) mixed with bran and/or lupin flour, although informants from Crete expressed surprise at such use of pressing waste. It was reckoned in Messinian Palaio Loutro that a household could stall-feed only one or two of its 5–20 pigs, even on bran mixed with olive pressings, and the remainder were taken to forage in the fields and woods.

Growth Rate

Enclosed wild boar crosses at Exokhi achieve dressed carcasses of 40–50kg by 10–11 months, but are fed year-round, and quite intensively during weaning and the following weeks. Stavros' pigs of traditional breed at Megali Panagia, which similarly achieved dressed weights of 35–50kg by 9 months, were fed little, but ranged freely in acorn- and mast-rich woodland in the two months preceding slaughter at Christmas. The enclosed wild boar on Mt Olympus are fed lightly and do not achieve dressed carcasses of 50kg until 1.5–2 years. Similarly slow growth is reported for pigs of wild stock, fed lightly and only in summer, at Ano Milia in the Pieria Mountains. In all these herds, most males are castrated young (at 1–2 months on Mt. Olympus, 3–4 months by one herder in Khalkidiki) and, by 1–2 years, are similar in weight to females, but may have more fat (unlike *breeding* males). Differences between herds, in mobility, quality of forage and level of feeding, make comparison difficult and the norms cited conceal considerable variability between seasons, years and individual pigs. These figures, however, support the contention of herders that animals of wild stock put on weight significantly more slowly than their domestic counterparts. Pigs of wild stock also develop slowly in terms of fat: at the Mt. Olympus farm, the skin and subcutaneous fat of a recently slaughtered adult male weighed 70kg, compared with only 40kg of meat, but young pigs slaughtered at 1–2 years are lean. Indeed, it is the new demand for lean meat, as well as game, that makes these animals economical to rear – despite slow growth.

In the case of household pigs, more or less heavy feeding and restriction of movement for the last 2–3 months were the norm. Such *threftá* or *threftária* ('fed animals') are reported as achieving, by 9–12 months, dressed carcasses of 70–100kg and sometimes 150kg. They were fed heavily to put on fat, sometimes to the point that their legs could not support them. The modest carcass weights of household pigs (50–60kg) often reported in Crete (*e.g.* Kalo Khorio, Partira, Pakhia Ammos, Vasiliki, Mirtos), where barley bread was normal, are attributed to the scarcity of grain, even for human consumption, as much as to the availability of olive oil as an alternative to lard (below).

Consumption

Today, extensively managed herds in Khalkidiki supply lean yearlings (9–12 months old, born in spring; 12–15 months, born in autumn) to local butchers and to urban customers who preserve pork, using freezers instead of traditional methods. The farm in Exokhi sells similar animals to a supermarket chain and that on Mt. Olympus supplies a gourmet restaurant. A few decades ago, levels of affluence and meat consumption were far lower, much rural demand was met by domestic production, and rural producers had poor transport links to urban markets.

Residents of Asoutaina in the 1950s fattened autumn piglets in the woods for sale in the regional capital at Christmas. At Aloides in the 1930s, Sofia's father reared and sold 20 pigs to buy barley flour in the town of Iraklio. At Yeni Gave, Manolis' family have for three generations roasted their own piglets, 4–6 weeks old (with dressed carcass of 12–15kg nowadays, 7–8kg 'in the old days') for travellers between central and western Crete. At Messinian Potamia in the 1930s, Panayotis' father kept one sow that produced two litters a year, from which he fattened one piglet for the house and sold the rest for fattening. He presumably killed the sows young as he fattened them for the autumn fair in nearby Khora, selling the meat as 'roast piglet'. In Paliambela-Kolindrou, herders provided some weaned piglets for fattening by neighbours and roasted the odd piglet at home ('2–4 piglets of 15kg each year for a household of twelve'), but sold many at seven months, thanks to proximity to the railway between Thessaloniki and Athens.

At Exokhi in the mid-twentieth century, however, Giorgos lacked easy access to urban market or main road and sold most piglets at 45 days, newly weaned, for neighbours to fatten. Most neighbours reared a household pig and several had breeding sows, so there was no market for yearlings unless a *kafenío* wanted one to roast. He retained the odd animal as replacement breeding stock or to fatten for Christmas, but if an adult sow was replaced she was fattened instead and produced more and better meat for less feed than a yearling because she already had a big frame. In Assiros, Vasilis sold a few yearlings in winter, but many villagers fattened their own pig (often one of his weaned piglets) and, without refrigerators, local butchers only occasionally sold pork. Even for weddings, a sheep or goat or occasionally a young piglet was slaughtered, but not a larger pig. Although Assiros was large and well connected to Thessaloniki, Vasilis abandoned pig farming because it was difficult to sell finished animals.

Fattened household pigs were slaughtered during the cool months, when meat was easiest to preserve. In summer, flies were a severe hazard and the carcass was at their mercy for longer because it was difficult to butcher until it had cooled down and the muscles had stiffened somewhat. In most regions, the peak of slaughter was in the days before Christmas, although the feast of St. Dimitrios (October 26th) was also popular near Thessaloniki (of which Dimitrios is the patron saint) and Carnival (February or March) was preferred in Messinia. Late slaughter in Messinia fitted with the practice of fattening pigs on olive pressings (olives were harvested in November–December and left to mature before pressing). The slaughter of the pig was a social occasion – widely referred to as *gourounokhará* or pig festival. Male relatives and neighbours gathered to kill the pig, gut and dress the carcass, shave or scorch its bristles, drink, and eat liver and other offal (fried by wives and daughters). As the saying went in Thessaly, 'you reared me like your child, but you slaughter me like your enemy' (Tziamourtas 1998, 247).

Adult pigs were sometimes skinned, with a knife (sheep and goats were flayed mostly with bare hands), and the skin was used for shoes or clogs. Younger pigs were not skinned and poor families might not skin an adult, to stretch their meat supply. The head and feet were skinned or scorched, and boiled until the meat came away from the bones. In northern Greece, a thick soup (*patsás*) was made, typically flavoured with salt, garlic and vinegar; on Crete, a gelatinous *pikhtí* or *tsiladiá*, that kept longer, was flavoured with salt, pepper, cumin, lemons and oranges. In Messinia, offal other than liver was chopped up, mixed with cooked cracked wheat (*pligoúri*) and herbs such as fennel, and stuffed into the large intestine to make *omatiés* (*cf.* Cretan *omathiés* – Psilakis and Psilakis 2001, 67–71; *matiés* in Tharounia) eaten more or less immediately.

Most lean meat was preserved. In northern Greece, some might be salted heavily and stored in a box or jar as *pastó* or *pastourmás* (it needed washing to be edible); some was chopped up with fat, perhaps flavoured with leeks or garlic, and stuffed into sausages that were hung to dry or boiled and stored in fat; and some was filleted, boiled with salt, pepper and perhaps bay leaves, then sautéed in fat to drive out moisture, and finally stored in a jar or box, sealed with fat, as *kavourmás*. This last method worked well, if the layer of fat was thick, and was also used for cast ewes, though these sometimes needed additional pig fat. On Crete, strips of lean meat (variously from the tenderloin, rump and shoulder) were steeped for a few days in vinegar and then suspended for a few days in sage smoke to make *apáki*; sausages (perhaps flavoured with rosemary, oregano or sage) were cured in a similar fashion; hams (sometimes salted) were smoked for rather longer (*hiroméri*); and meat from the lower limbs and around the ribs might be cooked and sealed in fat (*tsíglina*) in the same way as *kavourmás*. At Tharounia, much of the pig (including ham and shoulder) was again boiled and sealed in fat, but strips (*loúri*) from the breast were hung up indoors and preserved by drying and smoking. In Messinia, sausages and some lean meat were flavoured (with oranges, lemons, cinnamon, cloves, wine) and hung over aromatic smoke, but much of the carcass was scored and well salted, dried in a sack or basket for a week, then boiled with spices and oranges, strained and filleted, sautéed and sealed under fat (perhaps topped up with olive oil) in a ceramic jar as *pastó* (in northern Greece, meat preserved dry in salt). Other regional variants in Greek *charcuterie* are described by Kochilas (2001; also Andilios 2002).

On Crete, where most informants regarded vinegar and smoking as the main means of preserving meat, with salt added just to taste, sausages and *apáki* are usually said to have lasted a few months, from Christmas to March or April 'because, by May, it is hot and there are flies'. In Messinia and northern Greece, where salt was sometimes applied much more heavily (rules of thumb in Khora were 1kg of salt per 10kg of meat, or the weight of the head for a whole carcass), *pastó* and *kavourmás* are often said to have lasted a year – if not eaten sooner. Pork was perhaps expected to keep for less time in Crete, because household pigs were often smaller and the meat eaten more quickly than on the mainland. In practice, preservation methods depended on individual taste and the amount and quality (especially fat content) of meat to be stored (and hence on the capacity for fattening the pig), as well as on local tradition and storage conditions.

The bones of the household pig were used both fresh and after storage. For example, in Paliambela-Kolindrou, filleted leg bones might be broken and boiled with the *kavourmás* or salted and, over the following months, either cracked open and sucked as a marrow snack or used (as were salted ribs) to flavour vegetable dishes. In the same area, body fat and perhaps the thin layer of fat around the ribs were melted to *lígda* and used to seal *kavourmás* or stored in jars, while the thicker subcutaneous deposits, especially the layer (up to 10cm thick) along the back of the pig, were sliced off, salted, and stored as *lardí*. Both *lígda* and *lardí* could be stored for a year or more. In lowland southern Greece, olive oil was an alternative, but fat from the household pig remained important. In Messinian Khora, Khariklia boiled in milk the fat around the kidneys to make soft *vasilikó*, salted and stored in jars for use in pies and on pasta. The remaining fat, including that on the back, was chopped up and added to sausages or boiled to seal the *pastó* salt pork. The latter (with or without the pork) enhanced the flavour of a range of otherwise vegetarian dishes. Food was relatively plentiful in this household and they reared household pigs of modest size, knowing they could cook with olive oil when the tastier *pastó* ran out. In neighbouring Metaxada, Barba Ilias owned olive trees and a small press, but used olive oil only on salads and cooked with fat.

Meat from the household pig, preserved by drying, smoking, salting, pickling, and sealing in fat, was consumed over the following weeks or months, providing dietary diversity for the family on Sundays and, as was repeatedly emphasised, a ready source of *mezé* to welcome guests. Salted bones or fat or even meat routinely flavoured bean soups and casseroles, while strips of fat were taken as 'packed lunches' to the harvest or were heated to yield soft fat (and the discarded 'rinds' eaten with wine or spirits like 'pork scratchings' in British pubs). Soft fat was used to seal cooked pork and sometimes sausages, as an ingredient in savoury and sweet pies, and as the medium for frying – even to some extent in Crete, now renowned for culinary use of olive oil. These multiple uses of fat account for the widespread premium on heavy feeding of household pigs.

Strategies of Pig Husbandry

Recent management was somewhat polarised between rearing more or less substantial herds (a handful to a few tens of adult sows, one or more breeding males, and tens or hundreds of young pigs) and fattening one or two yearlings for household storage. Some informants pursued an intermediate strategy, keeping one or two breeding sows and selling a few piglets, and a substantial herd could be built up quickly from the odd household pig. Nonetheless, diet, fecundity and growth rate together encouraged a dichotomy between pig *breeding* and pig *fattening*. First, stall-fattening a yearling pig is roughly equivalent to feeding an additional adult human, and few households could rear more than one or two such animals per year. Secondly, pigs rapidly grow too big to be eaten fresh even by a large household and so tend either to be eaten/sold as piglets or, if slaughtered after a few months, to be preserved for storage or shared among many consumers (usually at a restaurant or fair, or through a butcher). Thirdly, pigs bear large litters and even a single young sow producing one litter of 3–4 piglets might meet the capacity for domestic consumption; every informant who owned a sow sold some young pigs (although need for cash was often a motive).

None of the herders we met had bred pigs throughout their adult lives and most had done so part-time, usually alongside arable farming. Vasilis in Assiros built up a herd to support his family, but Grigoris in Metaxada only kept sows and sold piglets when he needed dowries for his daughters. Some herders in Khalkidiki similarly bred pigs for a few years to finance building a house. Sofia's father in Aloides reared pigs, to bring in cash for buying flour, but only for a few years, perhaps because his children were old enough to collect acorns. Yorgos in Exokhi started rearing pigs as a teenager, to make use of his labour. Herd growth can be rapid: the herd on Mt. Olympus, despite harsh winters and limited success with second litters, grew in four years from an initial eleven sows and one boar to perhaps 300 pigs while providing young females and males for consumption. A flock of sheep, starting from eleven ewes and one ram, might have grown to one tenth of this size over four years, but arable farmers with teenage sons and needing additional income built up a flock of sheep far more commonly than a herd of pigs. Sheep produced milk (cheese was easier to sell than meat) and wool (sold or used by the household), whereas pigs only produced piglets, the market for which was easily saturated.

Barba Mitsos, the oldest resident of Paliambela-Kolindrou, says that 'in the old days, the pig was essential for the household', providing meat from Christmas to March or April and fat for the whole year. A large and reasonably well-off household might rear two or three such pigs, perhaps killing one in autumn, one at Christmas and one at Carnival, depending on the need for meat (*i.e.* number of visitors) and ability to provide fodder (one piglet might be killed early if they could not all be fattened). In different regions of Greece, beef, goat meat and especially mutton were also preserved in winter, usually by salting, but pork was normally preferred because the abundant fat made storage more reliable. Some households kept a breeding sow and sold a few weaned piglets, as well as fattening the odd one. The sow might be slaughtered after two or three litters and so play a dual role as breeding animal and then fattened *threftári*. Pigs produce large litters, however, and most households acquired a piglet from a neighbour with a sow or from the owner of a herd. Many bought a replacement piglet as soon as they slaughtered the previous one, but some waited a few months because they only needed, or could only feed, a pig of modest size.

Discussion: Greek Pig Husbandry in Context

Pig husbandry in Greece exhibits considerable diversity in scale (from household fattening to breeding large herds), in control over movement (from close confinement to free-range) and hence over diet and reproduction, and in the age and size at which animals are slaughtered or sold (mainly weaned piglets for fattening or finished pigs 1–2 years old). Husbandry regimes are similarly diverse, and often for similar reasons (human population density, extent of cultivation, consumption within or beyond the household), in the West Mediterranean (*e.g.* Albarella *et al.* 2007; Parsons 1962) and highland New Guinea (*e.g.* Rosman and Rubel 1989; Brown 1978, 87–94). Some detailed parallels in management can plausibly be attributed to the shared biology and ethology of European and Oceanian pigs and to the ecological similarities between Greece and the West Mediterranean. For example:

- breeding males are slaughtered young and others castrated at a few weeks in all three regions (Albarella *et al.* 2007, 295 table 16.1; Santamariña 1985, 307; Rosman and Rubel 1989, 28–29; Sillitoe 2007);
- the diet of extensively managed pigs follows a similar seasonal rhythm in Greece and the West Mediterranean (Parsons 1962, 229), while small numbers of pigs are widely fattened on food leftovers/by-products and spoiled/sub-standard crops in all three regions (Santamariña 1985, 308; Rappaport 1968, 59–61; Rosman and Rubel 1989, 29; Sillitoe 2007, 346);
- as in Greece, to limit damage to crops and ensuing social conflict, foraging pigs in the West Mediterranean may have their snouts wired (Albarella *et al.* 2007,

304 fig. 16.12; Santamariña 1985, 307; authors' field notes – Zureda, Asturias) and in New Guinea may be hobbled or tethered (Brown 1978, 90; Rappaport 1968, 161);
- free-range domestic pigs are enticed home with food, and genetic mixing occurs between domestic and wild/feral populations, in all three regions (Albarella *et al.* 2007; Rosman and Rubel 1989);
- small numbers of pigs may be reared and fattened like pets in all three regions (authors' field notes – Tiós, Asturias; Rosman and Rubel 1989), even to the extent that the meat is not eaten within the household (Rosman and Rubel 1989, 31).

Similarities and contrasts in *consumption* of pigs are fascinating. Across the Christian Mediterranean, the capacity to accumulate fat traditionally made pigs the preferred source of stored meat and fat (*cf.* Comet 1993; Grieco 1993). Strong regional traditions in preparation for storage included reliance on fat or salt in northern Greece, extensive use of herbs, spices and citrus fruits in southern Greece, and a combination of vinegar and smoking in Crete, while smoking was important in Italy and Spain. These traditions are often key elements in regional culinary identity and as such are normative ideals, from which individual preference and practical circumstances may favour divergence. Meat of hunted pigs is smoked in New Guinea (Rosman and Rubel 1989, 32), but the European emphasis on preserving pork gives way to slaughter of domestic pigs on ceremonial occasions when carcasses are distributed for relatively rapid consumption (*e.g.* Brown 1978, 91–94; Rappaport 1968, 81–84, 213–214). Distribution, rather than storage, of meat in highland New Guinea must be understood in its *social* context: residence may not be family based and feasting is central to interaction within and between communities (*e.g.* Brown 1978, 47, 59; Sillitoe 2007, 354–355). The contrast with Europe should not be overstated, however: informants in Greece, Italy and Spain (also Santamariña 1985, 322) emphasise that preserved pork not only enhanced household diet, but enabled hospitality to *unexpected* visitors or those offering labour. Such hospitality displayed a household's economic strength and so helped forge and maintain relationships with in-laws, exchange partners, patrons and clients.

Archaeological Implications

Potential insights from recent practice in Greece are considered briefly for four aspects of past exploitation of pigs: domestication and its bioarchaeological recognition; seasonality of diet, reproduction and slaughter; scales of management; and social context(s) of consumption.

Domestication and its Bioarchaeological Recognition

The variable control exercised in recent Greek pig husbandry underlines the difficulty of drawing a clear distinction between domestic and wild forms of *management* (Higgs and Jarman 1969; Rosman and Rubel 1989; Dwyer 1996; Albarella *et al.* 2006). For example, free-range 'domestic' pigs, that are enticed home by modest amounts of food, differ from wild pigs, that raid growing crops, in the archaeologically intangible sense that the former are fed intentionally by their owner (*cf.* Ingold 1986, 113) and the latter unintentionally by someone who has rights over them only if s/he succeeds in catching or killing them. Moreover, considerable *genetic* flow occurs between 'domestic' (owned) and 'wild' (free) populations: intentionally, when the latter are captured and enclosed or (as in New Guinea) when all domestic males are castrated and domestic sows can only mate with feral boars (Rosman and Rubel 1989; Dwyer 1996); and unintentionally, when wild boar mate with domestic sows or domestic males and females go feral. Consequently, genetic data shed no light on whether pigs are domestic or wild, in terms of *management* as opposed to *ancestry* – a point illustrated by a modern 'wild' boar from Italy with east Asian haplotype that is presumably a feral descendant of a recently introduced, improved domesticate (Larson *et al.* 2007, S1 Discussion).

A recent palaeogenetic study illustrates the complexity of the relationship between genotype and management. Ancient mitochondrial (maternally inherited) DNA indicates a consistent separation between Mesolithic (presumably 'wild') pigs from European and Near Eastern sites. At Neolithic sites in Romania, 'wild' (biometrically large) specimens were of European ancestry, but many 'domestic' (small) pigs were of Near Eastern haplotype and presumably descendants of introduced early domesticates (Larson *et al.* 2007). At Neolithic sites in central and western Europe, and later sites throughout the continent, 'domestic' pigs of European descent displaced those of Near Eastern type. Larson and colleagues (2007, 15279) infer *local domestication* consequent upon the introduction of Near Eastern domesticates, but caution that 'the process . . . (and possibly the degree of intention among early farmers) could have been fundamentally different from that in the Near East'. What does this mean in behavioural, rather than genetic, terms?

That wild females of European ancestry were added to managed livestock is unsurprising – in light of recent analogy, some were probably captured as piglets and others attracted by crop stands or refuse around settlements. But, assuming available samples are representative, how and why did pigs of local descent *replace* imported stock, perhaps over just 500–1000 years (Larson *et al.* 2007, 15280 table 2)? Some farmers perhaps favoured pigs of local descent for their appearance, but wholesale replacement on such grounds seems improbable. Conceivably, a European disease wiped out pigs of Near Eastern descent, but the timescale of replacement seems long for a catastrophic cause. Alternatively, trapping and voluntary recruitment perhaps led to replacement because pigs of Near Eastern descent were much fewer and/or less fertile than their local wild relatives. If domestic pigs were kept in small numbers, the risks of inbreeding and hence reduced fertility (or reduced viability of offspring) will have been high, especially given that small numbers are more easily kept under control sufficiently close to minimise interbreeding with wild boar (Dwyer 1996). Inbreeding is also likely to have been accentuated as pigs of Near Eastern descent passed through successive genetic bottlenecks on the expanding agricultural frontier. Interpretation of the changing genetic make up of domestic pigs, therefore, requires an understanding of management practices and population dynamics, based on biometric and morphological evidence for long-term selective pressures coupled with demographic, pathological, isotopic, bone geometry and histology, and dental microwear evidence of individual life histories (cf. Albarella et al. 2006; Dobney et al. 2006; Mainland et al. 2007; Ward and Mainland 1999). In conjunction with such complementary evidence, palaeogenetic data may fruitfully reveal how humans exploited pigs of varying ancestry, but in isolation they cannot resolve whether European wild boar were locally domesticated (cf. Zvelebil 1995) except in a narrow biological sense.

Seasonality of Diet, Reproduction and Slaughter

Seasonality of reproduction and slaughter is shaped by, and so may illuminate, strategies of pig husbandry and their ecological and social context (e.g. Vanpoucke et al. 2007). As in Corsica and Sardinia (Albarella et al. 2007), pigs farrow year-round in Greece, with a preference for spring accentuated by increasing latitude and altitude but reduced by feeding and housing (also Lauwerier 1983). The often gracile build of 'domestic' pigs (e.g. in later Neolithic and Bronze Age mainland Greece) perhaps argues against high levels of nutrition favouring regular, twice-yearly farrowing. Zooarchaeologists cannot, however, assume timing or frequency of farrowing, and so cannot infer season of death, without evidence (e.g. enamel hypoplasia – Dobney et al. 2007) for season of birth. Estimates of age at death, based on tooth *wear*, are probably also subject to greater error than in the case of ruminants, because the degree to which pigs dig for food (a major source of dental attrition) is very variable.

Extensively managed pigs dig in seasons and years when food above ground is sparse, but the earth soft. Pigs confined to a small pen rapidly remove herbaceous plants and may kill trees by stripping bark and digging until roots are exposed. Such seasonal, interannual and density-dependent variation in rooting may account for the very variable rate of tooth wear observed (by comparing adjacent teeth or anterior and posterior parts of the same tooth) in some faunal assemblages from Greece. Today, the leanest season is often summer, but in wet years summer annuals in stubble fields provide rich grazing for pigs and domestic ruminants. Some of these are C_4 plants (e.g. *Portulaca oleracea*), as are some whose roots pigs seek in winter as well as summer (e.g. *Cyperus longus*). Seasonal and interannual variability in local abundance of C_4 weeds and ruderal plants, therefore, may account for observed variability in carbon isotope values both of ruminant and non-ruminant fat residues in ceramics at Late Neolithic Makriyalos (Urem-Kotsou 2006, 219 and fig. 6.29) and of domestic animal bones (e.g. sheep at Neolithic Çatal Höyük – Richards et al. 2003, 70), without invoking long-distance herding or use of C_4 crops (e.g. common millet, *Panicum miliaceum*) as fodder.

Variability in pig diet poses challenges for zooarchaeologists, but also opportunities: given the linkage between habitat, nutrition and reproduction, integration of multiple lines of evidence can yield significant insights into husbandry. For example, at Roman-Early Byzantine Sagalassos in Turkey, enamel hypoplasia data imply a single farrowing in spring, an outcome favoured by high altitude but also suggestive of extensive herding, rather than household rearing which tends to dampen seasonality in diet and breeding. Two observed peaks of slaughter per year were perhaps shaped, therefore, by demands of consumption (e.g. seasonal festivals) rather than constraints of production (Vanpoucke et al. 2007). Either way, dental microwear suggests intensive fattening prior to slaughter (Vanpoucke et al. 2009).

Scales of Management

Late Bronze Age texts from the 'palaces' of southern Greece imply two scales of pig management. A few texts from Knossos and Pylos, recording tens or hundreds of pigs, imply management of large herds, at least at the point of mobilisation by the palace. Others from Pylos and Thebes list small numbers of fattened (*si*) or finished (*o-pa*) pigs, some at least destined for feasting or sacrifice. Such animals were provided, or fattened, by individuals such as the local leaders, each responsible for between two and six fattened pigs, on Pylos text Cn608 (Killen 1994; Halstead 2002; Bendall 2004; Palaima 2004). In the recent past, fattening of even two pigs was a significant undertaking and the generally gracile build of pig bones from Pylos (Halstead and Isaakidou research in progress) suggests that intensive feeding was unusual. For Roman Italy, textual, iconographic and biometric evidence similarly points to two scales of pig management, with extensive husbandry more common than intensive (MacKinnon 2001).

For the text-less Neolithic, recent management suggests avenues for zooarchaeological investigation of scale of husbandry. Declining size of domestic pigs through the Neolithic in *mainland* Greece (von den Driesch 1979, 14 fig. 5) implies relative reproductive isolation from wild boar. Moreover, at Late Neolithic Makriyalos, while variable rates of tooth wear (Halstead research in progress) suggest that pigs were free to root rather than confined in fattening pens, isotopic hints of dietary divergence from wild boar (Triantaphyllou 2001, 137 fig. 7.17) suggest that they foraged in the settlement or fields, rather than roaming in woodland, and thus that *aggregate* numbers were modest (Halstead 2006a, 45–46). Husbandry on a modest scale would have significant implications for population dynamics of domestic pigs and their genetic relationship with wild boar (above), for human impact on the landscape, and for the contribution of pigs to human diet (below). Intriguingly, size decrease is not evident at Neolithic-Bronze Age Knossos, raising the possibility of more extensive husbandry in the predator-free Cretan landscape, while some large pigs may indicate establishment of a feral population (Isaakidou 2004).

Whether Neolithic pigs were under community or household ownership is problematic, but architectural evidence arguably favours the latter (*e.g.* Halstead 2006b). As in the recent past, a single sow might easily have produced more offspring than a Neolithic household could consume, given that human skeletal evidence does not suggest a strongly carnivorous diet (*e.g.* Triantaphyllou 2001, 117–141; Papathanasiou 2003) and that many or most pig mandibles are from yearlings or older animals (*e.g.* Pefkakia – Jordan 1975; Ag. Sofia – von den Driesch and Enderle 1976; Knossos – Isaakidou 2004; Makriyalos – Halstead research in progress). Moreover, while recent households, that retained a young female long enough to produce one or two litters, might cater primarily for domestic consumption, some mature and old Neolithic sows perhaps imply breeding geared to provision of pigs or meat beyond the household. Organic residues in cooking pots from Neolithic northern Greece hint at variability *within* settlements in types of animal product (milk, meat of ruminants and non-ruminants) consumed, that is compatible with some specialisation in species of livestock reared by individual households (Urem-Kotsou 2006, 224; Urem-Kotsou and Kotsakis 2007, 239). *If a few households within a Neolithic community reared (and perhaps slaughtered) many of the pigs consumed, was this long-term specialisation or a short-term strategy to exploit surplus labour or grain and/or to prepare for a rite of passage celebrated with meat?* This is unanswerable without fine chronological resolution, although isotopic or dental microwear evidence for diet of animals from different contexts of consumption (*cf.* possible fattening of sheep for a 'feast' at Makriyalos – Mainland and Halstead 2005) could shed light.

Social Context(s) of Consumption

Pigs offer no secondary products (except manure) and so age of slaughter may be more informative of the form and context of meat consumption than with domestic ruminants. Neolithic pigs killed in their second year or later were arguably too big for fresh consumption by a household, even assuming the slow growth rate of recent wild stock, and so were presumably shared or exchanged (Halstead 2007), unless pork was regularly stored. Adipose fat, much of it from non-ruminants (presumably pigs), is prevalent in cooking pot residues (Urem-Kotsou 2006), but the modest size of these vessels (Urem-Kotsou 2006) does not favour large-scale preparation of pork for storage in fat (Halstead 2007). The size of butchery 'parcels' from Late and perhaps Middle Neolithic Greece, compared to that of cooking pots, suggests that domestic animals were cooked in ovens or pits and the bones subsequently broken to extract marrow (Isaakidou 2004; 2007; Halstead 2007), so ceramic residues may contain fat from this source, perhaps stored and added as flavouring to plant-based dishes as in recent cuisine. Bone deposition at Early Neolithic Paliambela-Kolindrou and Revenia-Korinou (Halstead and Isaakidou research in progress) and Early-Final Neolithic Knossos (Isaakidou 2004) suggests that butchered carcasses of pigs and other animals were distributed widely for consumption, in a manner more reminiscent of collective feasting in highland New Guinea than of domestic storage in the recent Mediterranean. Initial integration of faunal, isotopic and ceramic data thus suggests that Neolithic 'households' in Greece were significantly less closed, in terms of carnivorous commensality, than their recent rural counterparts (Halstead 2006b).

Conclusion

'Traditional' pig husbandry and consumption in rural Greece broadly parallel recent practice in the West Mediterranean, although detailed differences in preparation for storage helped shape regional culinary identities. There are also similarities with pig *husbandry* (but not consumption) in the ecologically and socially different context of highland New Guinea. In Greece and the West Mediterranean, husbandry and consumption are somewhat polarised between extensive herding for market and intensive fattening for domestic storage. These geographical similarities and contrasts, and the polarisation between scales of management, have heuristic value for zooarchaeological investigation. Recent practice highlights the complexity of the opposition between 'domestic' and 'wild' management and of its bioarchaeological recognition. Genetic exchange between domestic and wild populations is partly shaped by *scale* of management, which also influences seasonality of diet and reproduction, fertility levels and rates of growth. Such complexity, however, poses opportunities as well as problems for zooarchaeologists. Pig husbandry in the Neolithic of Greece may have fallen between the poles of recent practice and consumption patterns perhaps resemble collective feasting in New Guinea more than traditional storage for household-level hospitality in Greece. Consideration of recent husbandry in Greece, and of similarities and contrasts with other regions, has proved productive in thinking about pig management in the distant past, highlighting the value of recording traditional pig husbandry in Europe.

Acknowledgements

We are indebted to the patience of our informants, too many to name here, and to Manthos Besios and Yannis Stangidis who kindly introduced us to herders in Khalkidiki and Pieria, respectively. We also thank Umberto Albarella and an anonymous reviewer for helpfully critical comments on a draft of this paper.

References

Albarella, U., Dobney, K. and Rowley-Conwy, P. (2006) The domestication of the pig (*Sus scrofa*): new challenges and approaches. In M. A. Zeder, D. G. Bradley, E. Emshwiller and B. D. Smith (eds.) *Documenting Documentation: New Genetic and Archaeological Paradigms*, 209–227. Berkeley, University of California Press.

Albarella, U., Manconi, F., Vigne, J.-D. and Rowley-Conwy, P. (2007) Ethnoarchaeology of pig husbandry in Sardinia and Corsica. In U. Albarella, K. Dobney, A. Ervynck and P. Rowley-Conwy (eds.) *Pigs and Humans, 10,000 Years of Interaction*, 285–307. Oxford, Oxford University Press.

Andilios, N. (2002) Christmas in Cyprus: a traditional feast or a feast of traditions? In P. Lysaght (ed.) *Food and Celebration: from Fasting to Feasting*, 361–368. Ljubljana, ZRC Publishing.

Bendall, L. (2004) Fit for a king? Hierarchy, exclusion, aspiration and desire in the social structure of Mycenaean banqueting. In P. Halstead and J. Barrett (eds.) *Food, Cuisine and Society in Prehistoric Greece*, 105–135. Oxford, Oxbow.

Binford, L. R. (1978) *Nunamiut Ethnoarchaeology*. New York, Academic Press.

Brown, P. (1978) *Highland Peoples of New Guinea*. Cambridge, Cambridge University Press.

Comet, G. (1993) Le vin et l'huile en Provence médiévale, essai de bilan. In M.-C. Amouretti and J.-P. Brun (eds.) *La production du vin et de l'huile en Méditerranée*. Bulletin de correspondance hellénique supplementary volume 26, 343–358. Athens, École Française d'Athènes.

Dobney, K., Ervynck, A., Albarella, U. and Rowley-Conwy, P. (2007) The transition from wild boar to domestic pig in Eurasia, illustrated by a tooth developmental defect and biometrical data. In U. Albarella, K. Dobney, A. Ervynck and P. Rowley-Conwy (eds.) *Pigs and Humans, 10,000 Years of Interaction*, 57–82. Oxford, Oxford University Press.

Driesch, A. von den (1979) Haus- und Jagdtiere im vorgeschichtlichen Thessalien. *Prähistorische Zeitschrift* 62, 1–21.

Driesch, A. von den and Enderle, K. (1976) Die Tierreste aus der Agia Sofia-Magoula in Thessalien. In V. Milojcic, A. von den Driesch, K. Enderle, J. Milojcic-v. Zumbusch and K. Kilian *Die Deutschen Ausgrabungen auf Magulen um Larisa in Thessalien, 1966*, 15–54. Bonn, Rudolf Habelt.

Dwyer, P. D. (1996) Boars, barrows, and breeders: the reproductive status of domestic pig populations in mainland New Guinea. *Journal of Anthropological Research* 52, 481–500.

Grieco, A. (1993) Olive tree cultivation and the alimentary use of olive oil in late medieval Italy (ca 1300–1500). In M-C. Amouretti and J-P. Brun (eds.) *La production du vin et de l'huile en Méditerranée*. Bulletin de correspondance hellénique supplementary volume 26, 297–306. Athens, École Française d'Athènes.

Halstead, P. (2002) Texts, bones and herders: approaches to animal husbandry in Late Bronze Age Greece, *Minos* 33–34, 149–189.

Halstead, P. (2006a) Sheep in the garden: the integration of crop and livestock husbandry in early farming regimes of Greece and southern Europe. In D. Serjeantson and D. Field (eds.) *Animals in the Neolithic of Britain and Europe*, 42–55. Oxford, Oxbow.

Halstead, P. (2006b) *What's Ours is Mine? Village and Household in Early Farming Society in Greece*. G. H Kroon Memorial Lecture 28. Amsterdam, University of Amsterdam.

Halstead, P. (2007) Carcasses and commensality: investigating the social context of meat consumption in Neolithic and Early Bronze Age Greece. In C. Mee and J. Renard (eds.) *Cooking Up the Past: Food and Culinary Practices in the Neolithic and Bronze Age Aegean*, 25–48. Oxford, Oxbow.

Higgs, E. S. and Jarman, M. R. (1969) The origins of agriculture: a reconsideration. *Antiquity* 43, 31–41.

Ingold, T. (1986) *The Appropriation of Nature: Essays on Human Ecology and Social Relations*. Manchester, Manchester University Press.

Isaakidou, V. (2004) *Bones from the Labyrinth: Faunal Evidence for the Management and Consumption of Animals at Neolithic and Bronze Age Knossos, Crete*. Unpublished PhD thesis, University College London.

Isaakidou, V. (2007) Cooking in the labyrinth: exploring 'cuisine' at Bronze Age Knossos. In C. Mee and J. Renard (eds.) *Cooking Up the Past: Food and Culinary Practices in the Neolithic and Bronze Age Aegean*, 5–24. Oxford, Oxbow.

Jordan, B. (1975) *Tierknochenfunde aus der Magula Pevkakia in Thessalien*. Unpublished thesis, University of Munich.

Killen, J. T. (1994) Thebes sealings, Knossos tablets and Mycenaean state banquets. *Bulletin of the Institute of Classical Studies* 39, 67–84.

Kochilas, D. (2001) *The Glorious Foods of Greece: Traditional Recipes from the Islands, Cities, and Villages*. New York, William Morrow.

Larson, G., Albarella, U., Dobney, K., Rowley-Conwy, P., Schibler, J., Tresset, A., Vigne, J.-D., Edwards, C. J., Schlumbaum, A., Dinui, A., Balaçsescu, A., Dolman, G., Tagliacozzo, A., Manaseryan, N., Miracle, P., Van Wijngaarden-Bakker, L., Masseti, M., Bradley, D. G. and Cooper, A. (2007) Ancient DNA, pig domestication, and the spread of the Neolithic into Europe. *Proceedings of the National Academy of Sciences* 104, 15276–15281.

Lauwerier, R. C. G. M. (1983) Pigs, piglets and determining the season of slaughtering, *Journal of Archaeological Science* 10, 483–488.

Loukopoulos, D. (1983) *Georgika tis Roumelis*. Athens, Dodoni.

MacKinnon, M. (2001) High on the hog: linking zooarchaeological, literary, and artistic data for pig breeds in Roman Italy. *American Journal of Archaeology* 105, 649–673.

Mainland, I. L. and Halstead, P. (2005) The diet and management of domestic sheep and goats at Neolithic Makriyalos. In J. Davies, M. Fabis, I. Mainland, M. Richards and R. Thomas (eds.) *Diet and Health in Past Animal Populations: Current Research and Future Directions*, 104–112. Oxford, Oxbow.

Mainland, I., Schutkowski, H. and Thomson, A. F. (2007) Macro- and micromorphological features of lifestyle differences in pigs and wild boar. *Anthropozoologica* 24, 89–106.

Palaima, T. (2004) Sacrificial feasting in the Linear B documents. In J. Wright (ed.) *The Mycenaean Feast*, 97–126. Princeton, American School of Classical Studies at Athens.

Papathanasiou, A. (2003) Stable isotope analysis in Neolithic Greece and possible implications on human health. *International Journal of Osteoarchaeology* 13, 314–324.

Parsons, J. J. (1962) The acorn-hog economy of the oak woodlands of southwestern Spain. *Geographical Review* 52, 211–235

Psilakis, M. and Psilakis, N. (2001) *Kritiki Paradosiaki Kouzina*. Iraklio, Karmanor.

Rappaport, R. A. (1968) *Pigs for the Ancestors*. London, Yale University Press.

Richards, M. P., Pearson, J. A., Molleson, T. I., Russell, N. and Martin, L. (2003) Stable isotope evidence of diet at Neolithic

Çatal Höyük, Turkey. *Journal of Archaeological Science* 30, 67–76.

Rosman A. and Rubel, P. G. (1989) Stalking the wild pig: hunting and horticulture in Papua New Guinea. In S. Kent (ed.) *Farmers as Hunters*, 27–36. Cambridge, Cambridge University Press.

Santamariña, J. A. F. (1985) Ganado porcino: modalidad de existencia y papel en el seno de una comunidad rural de Galicia. *Trabalhos de Anthropologia e Etnologia* 25, 297–325.

Sillitoe, P. (2007) Pigs in the New Guinea highlands: an ethnographic example. In U. Albarella, K. Dobney, A. Ervynck and P. Rowley-Conwy (eds.) *Pigs and Humans, 10,000 Years of Interaction*, 330–356. Oxford, Oxford University Press.

Triantaphyllou, S. (2001) *A Bioarchaeological Approach to Prehistoric Cemetery Populations from Central and Western Greek Macedonia*. BAR International Series 976. Oxford, Archaeopress.

Tziamourtas, Z. (1998) *Laografiki Pinakothiki ton Karagounidon (Ilikos kai Pneumatikos Vios)*. Karditsa, Nomarkhiaki Autodioikisi.

Urem-Kotsou, D. (2006) *Neolithiki Keramiki tou Makrigialou: Diatrofikes Sunithies kai oi Koinonikes Diastasis tis Keramikis*. Unpublished PhD thesis, Aristotelio Panepistimio Thessalonikis.

Urem-Kotsou, D. and Kotsakis, K. (2007) Pottery, cuisine and community in the Neolithic of north Greece. In C. Mee and J. Renard (eds.) *Cooking Up the Past: Food and Culinary Practices in the Neolithic and Bronze Age Aegean*, 225–246. Oxford, Oxbow.

Vanpoucke, S., de Cupere, B. and Waelkens, M. (2007) Economic and ecological reconstruction at the Classical site of Sagalassos, Turkey, using pig teeth. In U. Albarella, K. Dobney, A. Ervynck and P. Rowley-Conwy (eds.) *Pigs and Humans, 10,000 Years of Interaction*, 269–282. Oxford, Oxford University Press.

Vanpoucke, S., Mainland, I., de Cupere, B. and Waelkens, M. (2009) Dental microwear study of pigs from the classical site of Sagalassos (SW Turkey) as an aid for the reconstruction of husbandry practices in ancient times. *Environmental Archaeology* 14, 137–154.

Ward, J. and Mainland, I. L. (1999) Microwear in modern free-ranging and stall-fed pigs: the potential of dental microwear analysis for exploring pig diet and management in the past. *Environmental Archaeology* 4, 25–32.

Zvelebil, M. (1995) Hunting, gathering, or husbandry? Management of food resources by the Late Mesolithic communities of temperate Europe. *MASCA Research Papers in Science and Archaeology* 12, 79–104.